Praise fo

"If you love animals, you'll lo ⎯⎯⎯ ⎯⎯⎯ combines up-to-date knowledge about animal communication with Friend's clear writing and compelling fascination with animals and the people who study them."

—Patricia B. McConnell, Ph.D., author of *The Other End of the Leash: Why We Do What We Do Around Dogs*

"In a very enjoyable and well-informed book, Tim Friend argues that modern science is surpassing Dr. Dolittle. We know enough about the communication of other species to conclude, with the author, that animals may not have language, but they have plenty to say."

—Frans de Waal, author of *The Ape and the Sushi Master*

"In Tim Friend, the animal kingdom—and for that matter, the plant and bacterial kingdoms—has found a thoroughly engaging new spokesman. *Animal Talk* makes a bold case for the argument that the richness of animal communication systems differs from human language only in degree, not in kind. This is a must-read for anyone who wants to listen in on what the other inhabitants of Planet Earth have to say to each other and what they can teach us as well."

—James Shreeve, author of *The Genome War* and *The Neanderthal Enigma* (Coauthor with Donald Johanson of *Lucy's Child*)

"Tim Friend has just created something of a Rosetta stone for understanding what creatures are saying, in his new book, *Animal Talk*."

—*The Boston Globe*

"Friend has discovered that his newfound knowledge can be applied to human interactions. He's given better dating advice to friends . . . and correctly predicted the outcome of the Iowa caucuses to a National Public Radio producer." —*The Boston Globe*

"A comprehensive and engaging overview." —*Science News*

Animal Talk is "generous with scientific detail, but keeps it light with first-person anecdotes and helpings of humor."
—*Psychology Today*

"A delightful, entertaining, and instructive book for the general public on animal communication. All kinds of animal lovers will enjoy his presentation of 'the bigger picture' of animal talk in the wild. . . . Friend is an engaging writer; throughout, he displays a deft ability to capture the world of animal sound with just the right phrase." —*Publishers Weekly*

"Science journalist Friend strikes a mother lode of popular interest. . . . Readers will be most intrigued by the unheralded figures Friend found, such as a 'real-life [Dr.] Dolittle' . . . Accessible prose . . . Animal lovers are sure to be enthralled." —*Booklist*

"An amiable, anecdotally rich tour of communication . . . fresh with the latest ideas behind how and why we all send signals . . . His text boasts its own communicative clarity . . . A very credible overview."
—*Kirkus Reviews*

"A fun read . . . packed with clever experiments, intriguing anecdotes, and a delight in the diversity of animal behavior." —*Discover*

ANIMAL TALK

Breaking the Codes of Animal Language

TIM FRIEND

FREE PRESS

New York   London   Toronto   Sydney

*f*P
FREE PRESS
A Division of Simon & Schuster, Inc.
1230 Avenue of the Americas
New York, NY 10020

For information regarding special discounts for bulk purchases,
please contact Simon & Schuster Special Sales at
1-800-456-6798 or business@simonandschuster.com

Book design by Ellen R. Sasahara
Graphics on pp. 17 and 148 redrawn by Mark Stein

Manufactured in the United States of America

1   3   5   7   9   10   8   6   4   2

The Library of Congress has cataloged the hardcover edition as follows:
Friend, Tim.
Animal talk : breaking the codes of animal language / Tim Friend.
p.   cm.
Includes bibliographical references (p.) and index.
1. Animal communication. I. Title.
QL776 .F65 2004
591.59—dc22   2003063107
ISBN 0-7432-0157-4
0-7432-0158-2 (Pbk)
Photo and Illustration Credits:
Black howler monkey, p. 56: Photo by Kevin Shafer/Programme for
Belize. Sound response using the DTAG, p. 148: Graphic by M. John-
son, Woods Hole Oceanographic Institution, from data collected under
NMFS permit #31014 and Canadian permit from the Dept of Fisheries
and Oceans #2000-489. Yellow-naped parrot, p. 156: © 1992 James
Gilardi, World Parrot Trust. Adult male lazuli bunting, p. 171: Photo by
Susan Riddle. Yearling male lazuli bunting, p. 172: Photo by Susan Riddle. Bowerbirds, p. 181: The Department of Special Collections, Kenneth Spencer Research Library, University of Kansas. Speckled wood
butterfly, p. 197: © Dr. Malcome W. Storey. Common dolphins, p. 213:
Courtesy Programme for Belize. Kanzi, the bonobo, p. 246: Photos
courtesy of Georgia State University Language Center.

Acknowledgments

THIS BOOK BEGAN several years ago with the suggestion from my soon-to-be agent, Jennifer Gates, that I try writing a book about how animals communicate. My first thought was, How hard can it be to explain why a bird chirps? Animal communication turned out to be an incredibly difficult subject to try to learn. One really should go to school for this kind of thing. So, I am forever indebted to some very smart people who have dedicated their lives to truly understanding why a bird chirps, or a dog barks or, a sage grouse shakes its booty.

My conversations about animals began with Harold Gouzoules of Emory University, who explained the basics and directed me to other experts in the field. Thank you Peter Tyack at the Woods Hole Oceanographic Institution for allowing me to spend time hanging around clueless in your lab while you furthered my limited education about animal signals. Your files, which you so graciously allowed me to copy, provided the background and truly jump-started this journey. Vincent Janik, while working in Tyack's lab, patiently explained the importance of the mimicry of sound in animals and his cautious approach to interpreting data helped me to learn at least a little restraint as a lay person. Patricia McConnell, of the University of Wisconsin, spent hours guiding me out of an initial disillusionment with the whole topic and supported the idea that there is a universal form of communication in animals. Watching your border collies herd sheep at your farm was a real treat, not to mention the surreal dog show.

I am especially indebted to Lee Dugatkin, at the University of Kentucky, Eugene Morton, now at York University in Toronto, Jack Bradbury of Cornell University, and James Gould at Princeton University. Dugatkin, Morton, and Bradbury's textbook, and the hours upon hours spent talking and reading my early drafts, were more helpful than I can express. I hope the single malts helped ease the pain. Gould's work on mate choice and his guidance on sexual selection was critical.

Sherman Wilcox at the University of New Mexico and Talbot Taylor at the College of William and Mary, thank you for helping me understand your work on language. You are the only linguists whose words have ever made any sense to me. Thanks to all of the scientists who granted the many interviews that were conducted over the past couple of years, including Bernd Heinrich, Katy Payne, Roger Payne, Cynthia Moss, Marc Hauser, and Donald Griffin. Marc Bekoff deserves special credit for braving the new world of animal emotions, making it safe for wizards to come out from behind the curtain and for hacks like me to stumble along behind. Jane Goodall, thank you for allowing me to spend time with you, and for inspiring so many people.

If awards can be granted for patience, Leslie Meredith and Dorothy Robinson should have gold medals. Leslie, especially, thank you for your guidance and kindness. To Lynne Camoosa, thank you for reading the manuscript and correcting my dyslexic errors, and for never complaining about my absence.

For Kelsey, Amber, and Mom

Contents

One
A Walk in the Park: Toward a Universal Language

Two
Of Ravens and Robots 33

Three
The Show Must Go On 62

Four
One Language and Few Words 84

Five
The Chemistry of Love 112

Six
Songs and Shouts 138

Seven
Flash and Dance 165

Eight
Our House 192

Nine
All in the Family 212

Notes 251
Index 264

ANIMAL
TALK

A WALK IN THE PARK: TOWARD A
UNIVERSAL LANGUAGE

UST BEFORE MIDNIGHT, I crawl off my hammock and slip quietly from under the blessed drapes of mosquito netting. The kerosene lanterns have been dimmed to barely a flicker across the long raised wooden platform that serves as our base camp in the tropical rain forest of northwestern Peru. The platform is in the center of a clearing next to the bank of a small tributary that feeds into the Napo River half a mile away. I arrived with four other journalists and a guide late in the afternoon after hiking through the muddy and tangled jungle since early morning. We have joined two scientists, a few camp cooks, and the drunken pilot of a Cessna seaplane, who arrived just before dinner. The pilot is supposed to give us a flyover of the region first thing after breakfast. As much Johnny Walker Red scotch as he was putting away tonight, he'll still be drunk when we take off on the river early in the morning. The bush pilot's creed is, If you can't fly drunk, you can't fly.

This is my first trip to the Amazon rain forest. The closest I have come to a jungle until now is the tangled thicket of the Ozarks, where I grew up. Surprisingly, there are a lot of similarities, especially in the number of things that bite, sting, scratch, and burn. At the moment, the others in the expedition party are snoring. But that isn't why I'm awake. There's a raucous party going on—with lots of wild action by the strange and wondrous creatures all around our campsite. The night is teeming with sounds that seem louder and more intriguing than anything I have ever heard.

A minute ago, high in the canopy just beyond the camp's perimeter, something big crashed through the leaves. The only creature large enough around here to make such noise is a sloth. True to its name, it does not seem to be in much of a hurry. Bats have been fluttering since dusk through the rafters of the thatched roof above the platform. Earlier this evening, one swooped down and snatched a tarantula that was crawling up a colleague's mosquito netting. We could hear a soft crunch as the bat caught the spider with its teeth and darted off with its dinner into the night.

Beyond the camp, in vibrant surround sound, tree frogs and insects are laying down a soulful, energetic chorus like a choir at an old-fashioned Southern tent revival. Unfamiliar birds and nocturnal monkeys overlay the chorus with melodies and their own unique lyrics. This is one party I am not going to miss despite the rather condescending warning of the scientists not to leave camp alone at night. We could get lost *or worse,* they cautioned. I spend quite a bit of time in the field, but no matter where I travel, scientists tend to treat journalists like bad children who need constant supervision. Their admonishment only heightens my resolve to sneak out of camp.

The main attraction of this remote spot in the jungle is a canopy walkway constructed with ladders and suspension bridges that leads 115 feet straight up to the tops of the trees. Ordinarily, getting to the upper canopy entails climbing with harnesses and ropes. This is no simple task and usually involves close encounters of the unpleasant kind with nasty things that sting, burn, and bite. The canopy walkway, which bears a strong resemblance to the Swiss family Robinson's tree house, is a vast improvement on grappling and slapping one's way up. As far as I know, only a few of these walkways exist in the world. Tonight, this one is going to be my stairway to heaven.

I am on a mission and have a woefully short time to fulfill it: to learn how animals communicate with each other and what they spend so much time chattering to each other about. At this point, my quest seems absolutely overwhelming. Real experts devote entire careers to studying a single species of animal and are still left with many more questions than answers at the end of the day. My head is full of questions, too, which I plan to explore and explain in this book: If animal behavior is mostly instinctual, as scientists generally thought for more than a hundred years, why do animals need to communicate? If animals are thinking creatures and capable of emotions, as a growing number of scientists now believe, do their signals convey information (similar to our words)? Or are animals merely snarling or cooing to manipulate each other's behavior to get something they want (as we also often do)? How did the colorful, noisy, and smelly signals of the animal kingdom arise in the first place? Is any animal system of communication similar to human language? Do animals ever lie or attempt to deceive each other when communicating? Do the chirps, barks, and roars of different species have anything in common or follow predictable rules or patterns? Can a bird understand a monkey? Do species learn to communicate or is it all programmed by genes? To what extent is human communication, both verbal and nonverbal, programmed into our genes?

Scientists have been asking questions like these and working hard at finding the answers for more than a century, but there have been an enormous number of recent discoveries about animal communication. Studies on communication among tree frogs alone could fill a book. The eminent sociobiologist E. O. Wilson and the entomologist Bert Hölldobler produced a 732-page tome devoted to ants. I have three books in my home library on cichlid fishes, seven devoted to primates, five on dogs, more than a dozen on various species of birds. Most books focus on a single behavior, such as courtship rituals among birds, or the social behavior of primates, or the chemical signals of insects.

Yet surprisingly few books written for the general public have focused on the great range of animal communication. Usually, these books devote only a chapter or two to songs, dances, and scents. So my challenge here is to draw from the wealth of research conducted by hundreds of scientists and present the bigger picture of animal talk in the wild. The Amazon rain forest seemed like the best place to get a full immersion in nature and to begin eavesdropping on some animal conversations.

The few remaining unspoiled rain forests of the world are nature's Manhattan, London, and Tokyo—bustling organic metropolises with their own laws that govern every creature equally from conception through life and into death. The laws of nature demand procreation and a fight for survival, but the means developed to achieve those ends are tremendously varied. Mother Nature has fostered all manner of societies, cultures, learning, gaming, altruism, deception, cooperation, competition, industries, arms races, and intelligence. Look closely at any habitat and you can find daily dramas involving struggles between predators and prey, elaborate courtships, covert copulations, sibling rivalries, struggles for dominance, defense of territories, and many, many opportunities to arrive at a premature death. The same dramas are played out all over the world in every environment, from the deep ocean vents where microscopic life may have begun to the lawns and shrubs only a few steps away in the backyard.

Communication between all of the earth's creatures makes these dramas possible. Indeed, communication is the glue of animal societies. Without a means of communicating, no life, including the simplest single-celled organisms, could exist. Communication, like the tango, takes two. And it requires a signal, which can be anything from the release of chemicals between colonizing bacteria, to the come-hither flashes between male and female fireflies in the backyard, to the "let's go" rumble of African elephants, to the "signature" whistles of dolphins, to a dog barking simply to be let outside.

Over the course of our journey we will explore the origins of communication and how all of the marvelous signals employed by animals have developed. We will also look at why scientists think the way they do about animals. A pretty big divide separates scientists and laypeople, especially in their perspectives on animal behaviors. Yet, as ordinary people and pet owners become more informed and as scientists come to better appreciate the genuine intelligence of their animal subjects, both groups are moving toward a middle ground.

Before we head off into the jungle and climb the canopy walkway, we ought to know what scientists mean when they talk about animal communication. The basic textbook definition of animal communication taught to most undergraduate students states that it is "the provision of information by a sender to a receiver, and the subsequent use of that information by the receiver in deciding how to respond. The vehicle that provides the information is called the signal." One example of animal communication is the exchange between songbirds in a tree in the backyard. The male songbird is usually the sender. He sings his repertoire of songs, which is the signal, to a female songbird that has lighted on a nearby branch to listen. She is the receiver. The information would be something contained in the male's songs that helps the female decide whether the singer is a suitable mate. An exception to this male-dominated art and science is the northern cardinal—each sex sings to the other, and they even seem to duet. My mom's fat little Chihuahua, Taco, is another example. Taco has a habit of running to my mom and barking in a particular manner whenever my dad does not hang up his jacket when he comes home from work. Taco is the sender. Mom is the receiver, and the information is that dad left his jacket on the bed. (How my mom knows Taco's bark carries this specific meaning, however, is a mystery.)

E. O. Wilson takes the definition a bit further in *Sociobiology: The New Synthesis*. Wilson defines communication this way: "Biological communication is the action on the part of one organism (or cell) that alters the probability pattern of behavior in another organism (or cell) in a fashion adaptive to either one or both of the participants. By adaptive I mean that the signaling, or the response, or both, have been genetically programmed to some extent by natural selection. Communication is neither the signal by itself nor the response; it is instead the relationship between the two."

The exchange of chemical signals between bacteria is the oldest form of communication on the planet and provides a good example of Wilson's definition. For pathogenic bacteria to become harmful to us humans, they first need to reach a critical mass, which they do by communicating with each other essentially to take a head count. A single *E. coli* bacterium—the type that naturally lives in our gut but sometimes contaminates foods—will release a chemical signal that sends the message "I am here." If pathogenic bacteria are present, the message will cause them to release a similar signal that says "I'm here, too." If the bacteria sense that their numbers are strong enough to ward off an attack by the host's immune system, they will all respond by releasing their toxins into the host's cells. The signals are genetically programmed and create a dynamic relationship between the senders and receivers, a type of bacterial communication known as quorum sensing.

All types of chemical signals, including those used by animals to attract mates, are adaptive, by Wilson's definition. Chemical signals originated with the first group of bacteria that appeared on the young earth, about 3.8 billion years ago. Thanks to their highly effective signaling, bacteria often function as a type of superorganism and are one of the most successful forms of life on earth. They are also the only life-form that appears capable of living elsewhere in our solar system.

The signaling that allowed colonies of bacteria to thrive in all types of environments, from ocean vents to glacier ice, eventually gave rise to the cell-to-cell communication that made possible the evolution of multicelled organisms. This "adaptive" communication has facilitated the incredible success of the earth's insects, including the 8,800 species of ants, which Wilson and Hölldobler say constitute an amazing 15 percent of the earth's biomass. In fact, Wilson developed his definition of communication from the study of ants, which have evolved a complex social system by using specific chemical signals with unambiguous meanings. Any given chemical signal produced by a sender will elicit a specific, invariable response from a receiver. For example, if an invader enters a colony of fire ants, sentries will release a chemical alarm that summons other members of the colony to attack the invader. The members of an ant colony also carry a chemical badge that says to the others "I belong here." Spread a little of that chemical badge on an invader and the sentries will allow it to enter the colony as one of their own. The chemical signals used and understood by ants are obviously programmed into their genes.

The communication systems of all living creatures are programmed to some extent by genes. While genes rule supreme over insect communication, many mammals and birds have some flexibility for learning certain types of signals. Songbirds inherit a fixed genetic template of the songs they will sing as adults, but they must hear the songs of adults during a critical period of development to sing their songs correctly when they grow up. Some birds, such as cowbirds, can learn the songs of a different species. Cowbirds, known for laying their eggs in the nests of other birds, including golden warblers, are born with a genetic template for their own species' songs, but when raised in the nest of the golden warbler, they learn to sing the songs of their foster parent.

The development period of song learning among birds is similar to the babbling phase of human babies and other young primates, such as the vervet monkeys that live on the African savanna. Vervets are born with a genetic template for specific types of calls, but they must learn to use them in the proper contexts by observing adults. The foundation for human language—grammar and syntax—is genetically programmed in humans, but the ability to speak human language must be learned from exposure to adult speech. Many species of animals depend on experience and learning to communicate. The difficult part is figuring out which species need to go to "school" to communicate effectively and which are fully programmed by their genes. Humans are regarded by humans, of course, as the savants of communication. It may seem logical to conclude that the complexity of communication systems follows a hierarchy leading from humans to apes to mammals and birds and on down to insects. But it's not so simple. The most complex system of communication next to that of humans is found in the dance steps of the honeybee.

How can that be? Nature isn't concerned with how an organism communicates as long as it finds a way to do it successfully. The forces that help an organism achieve a successful system for communicating—whether simple or complex—are known as natural selection and sexual selection.

Natural selection basically means that nature favors any trait that improves an animal's chances of survival. If screaming a warning or releasing a certain chemical when a predator is sighted helps keep the family or kids from becoming dinner, then nature will favor those signals, which will become established in a species.

Sexual selection is an equally powerful force that shapes an animal's come-hither signals, such as lilting songs and flashy ornaments or displays. Sexual selection favors any signal that helps an animal win at the mating game. Whichever of the sexes is in the driver's seat when it comes to choosing a partner—usually females, as most scientists now realize—will have the greatest influence on shaping the signals that the opposite sex uses to woo a partner.

Some scientists argue that sexual selection has been more influential than natural selection in the evolution of communication. Step out onto your back porch on a spring morning or take an evening stroll on a country lane and it is easy to see why. At dawn and dusk, most of the animals and insects that can be heard are chatting about sex. The chirping of crickets, the croaking of bullfrogs, and the repertoires of songbirds are solicitations made by males at the insistence of incredibly critical and demanding female audiences. Likewise, the colorful plumage and exaggerated physical characteristics that many animals display have typically developed to attract members of the opposite sex.

The common definition of communication—that a sender provides information to a receiver through a signal—follows an "information model" to interpret the exchange. Some scientists argue that the information model reflects a human-centric bias because information is so highly valued in human society and is conveyed by human language. To understand better what many mammals and birds are experiencing when they communicate, it may be more useful to compare human nonverbal communication with the systems used by animals. From that perspective, communication becomes less about exchanging information per se, and more about managing or manipulating behavior.

In the late 1970s and 1980s, scientists Richard Dawkins and J. R. Krebs popularized the idea that communication is primarily a means of manipulating behavior. According to Dawkins, the bottom line for all species is to pass their genes to the next generation, assuring a type of genetic immortality. So each party in a conversation has its selfish interest at stake, and selfishness is at the root of all behavior. Dawkins later developed the notion of the "selfish gene," which underlies much scientific interpretation of animal behavior and communication today. Not everyone agrees with this interpretation of behavior, but Dawkins and Krebs are so commonly cited in scientific papers published on animal communication that their argument cannot be ignored. Most scientists would agree that communication is indeed motivated by the sender's need or desire to manipulate the behavior of another animal but that the receiver also extracts some key, albeit simple, information from a signal to make a response.

One question that scientists still debate is whether the sender or the receiver has the greater influence on the emergence and refinement of signals. Ornithologist Eugene Morton, of the Smithsonian Migratory Bird Center in Virginia, and Donald Owings, a professor at the University of California at Davis, both argue that the receiver, or "perceiver," as they call it, runs the show. They prefer the word *perceiver* because it indicates that the listener plays an active role. To describe their perspective on how vocal and visual signals have evolved, Morton uses the analogy of a stand-up comic. For example, Jerry Seinfeld, who is known to drop in unexpectedly at small comedy clubs around New York City to try out new material, drops into a club, delivers some new jokes, closely watches the audience, and sees that the jokes aren't going over well. Only a couple of people are chuckling, probably just because he's Jerry. He switches gears, moves to another set of jokes, and gets a slightly better response. He keeps trying new material until something clicks and he has the audience rolling with laughter. The bad jokes go into the trash and the good ones are kept for his road show. But while Seinfeld may make a hit with his new jokes, who is really in charge? Morton and Owings say it's the audience, because they determine which signals from the stage live or die.

In the animal kingdom, the sender must come up with a signal that gets the attention of the audience—the receiver—and elicits the desired response. If it does not, the force of natural selection or sexual selection will ensure that the unsuccessful sender fades from life's stage.

Last night, my group in the Amazon had an opportunity to witness an effective signal from a remarkably talented Peruvian guide, who had taken us on a night hike at another location in the rain forest. An hour into the hike, the guide suddenly thrust out his arms to block anyone from moving ahead and whispered in Spanish with unmistakable urgency, "Alto, alto!" He shone his flashlight in front of us, illuminating a juvenile fer-de-lance pit viper curled on a branch jutting out into the trail. These snakes, among the deadliest in the world, are on average five feet long and deliver about 105 milligrams of venom with a single bite—only 50 milligrams are needed to kill a person. The viper's bites are a common cause of premature death among people in Central and South America, particularly in the Amazon. Had one of us brushed against it, we would have had one person fewer in our party. The guide's urgent signal communicated danger and altered our behavior. If he had casually said, "Umm, there's a snake," we might not have paid much attention. All of the signals in the animal kingdom have proven effective over time at getting their intended audience's attention.

It is important to note here that a signal does not always require a response. Sometimes the receiver simply is not interested in whatever the sender has to say, no matter how effective the signal may be in other circumstances. Females that are being solicited by males for mating ignore the majority of the signals sent their way. Their lack of response can be interpreted as, "Ho hum. You are not quite what I'm looking for." This lack of response to mating signals certainly holds true for many human males, as well. Unreturned telephone calls and e-mails can be just as informative as a verbal reply such as, "I'd rather not be with you." Relationships with friends or lovers are over when the other person stops responding to calls or e-mails.

Sometimes it simply pays for the receiver to keep quiet. Harold Gouzoules, of Emory University in Atlanta, has studied primate behavior for more than two decades. Chimpanzees form social cliques that possess varying degrees of dominance and status. On any given day in the life of an average male chimpanzee, that male has a good chance of being pushed around or thumped on by a more dominant male, much like middle school for adolescent human primates. The usual response among chimpanzees to a beating is loud screaming, which is intended to solicit help from one's buddies. Sometimes that approach works, but it's not guaranteed. Gouzoules has found that the silent treatment from a chimpanzee's allies when it is being beaten by a more dominant male is a clear signal that says, "Uh, sorry, pal. It's not in my best interest at the moment to help you. If I look the other way I might win points with the big boy." On the other hand, if the male's allies respond to the plea for help and they win the conflict, their social group can boost their status in the community. The popular "reality" television show, *Survivor*, was nothing more than humans playing chimpanzee politics.

When communication occurs, it must take place through one or more of the animal's senses. Remarkably, nature has been quite conservative in the development of senses in its animal subjects. Although an estimated 10 million species currently reside on the planet, all must make use of five basic senses—vision, hearing, smell, touch, and, to a lesser degree, taste. The senses provide most animals with at least five basic types of signals to choose from: visual displays, vocal sounds, chemical signals, tactile signals, and sometimes messages sensed through taste. Overall, the environment determines the extent to which a particular sense is developed more than any other, or which senses might be combined in a particularly useful way. The five senses are well developed in humans, but animals possess senses that are often more acute or broader in their range. Some mammals, birds, and insects can hear sounds out of our range of hearing and see colors or patterns that are invisible to the human eye. Depending on the species, the environment, and the influences of natural and sexual selection, some senses will be favored over others. Ants rely most heavily on smell and touch. Moths rely primarily on uncanny olfactory abilities. Elephants can sense vibrations in the ground over long distances through the pads of their feet.

While most signals in the animal kingdom are limited to the five senses, nature has provided for two rather exotic exceptions. Out in the Amazon and in the streams beyond our camp are silvery knife fish that sing the body electric. Electric fields have been adopted as communication tools by a variety of aquatic species, including sharks, eels, and some species of fish, which use them to sense their prey and communicate with members of their own species. Electricity as a medium for signals appears to be limited to creatures that live in the water, which serves as a good conductor of electrical signals. The knife fish generate weak electrical charges to converse, defend territory, locate food, and navigate at night. When males send electric pulses to each other, they are usually saying, "My clump of plant. Find your own." But electrical charges take on an amorous tone between males and females. When knife fish mate, they swim in a spiraling ballet and sing an electric duet that rises and falls in intensity and cadence. Philip Stoddard, a specialist in electrical communication at Florida International University in Miami, recently learned that these couples can adjust the frequency of their electric love songs to keep one of their predators, the electric eel, from eavesdropping and devouring them.

Knife fish range from 3 to 12 inches long and thrive in tropical rivers and lakes. The eel that dines on these fish averages six feet long in size. It swims about attempting to eavesdrop on courting male and female knife fish and on squabbling, territorial males. The eel's sensitivity to the electrical signals of its prey is limited to low frequencies. The knife fish, which began its evolutionary journey with a simple signal at a low frequency, has developed a more complex pulse at frequencies above the eel's sensitivity. Stoddard made the discovery using an electric eel named Sparky and the tape-recorded signals of knife fish in a big tank. He put an electrode in the tank and released a simple signal that the eel could easily detect. The eel assumed it was dinner and responded by swimming to the electrode and blasting it with 300 to 400 volts, as it would to stun its prey. But when the knife fish's complex signals were played in the tank, the eel responded only a third as often, suggesting that the fish have altered their signal over time to avoid being eaten. The experiment also suggests that the eel may be catching on, since it can detect the complex signal a third of the time.

Electricity is a powerful force in a six-foot-long eel: the first time Stoddard conducted the experiment, he did not have his amplifier and sound equipment plugged into a surge protector. When Sparky became excited by the electrode that was mimicking the knife fish and gave it a big jolt, Stoddard's equipment was fried.

The other method of signaling that is unique among animals is known as bioluminescence, which is the ability to generate one's own light. Technically, this falls under the class of chemical signals, but generating light takes a species beyond the typical olfactory signals that account for most chemical communication. Bioluminescence is most widespread among marine animals that live in the deep ocean, although you can get a wink from a bioluminescent creature by standing on the beach at night and watching tiny dinoflagellates fluoresce in the roiling surf. Edie Widder of the Harborview Oceanographic Institution in Fort Pierce, Florida, argues that bioluminescence is the most underappreciated of all modes of communication. She estimates that it is used by 90 percent of all creatures that live in the darkness of the midocean, which ranges from about 500 feet to more than 3,000 feet deep. Sunlight does not penetrate below 500 feet in the ocean, so some animals there have developed their own headlamps and a few other truly bizarre features that employ light. Bioluminescent signals are given to attract mates, lure prey, and frighten potential predators. The deep ocean, which I have been fortunate to visit in the Russian submersible *Mir 1*, looks a bit like a surreal night sky with jellyfish, lantern fish, and other critters appearing as twinkling and shooting stars.

Of all the residents of the midocean, the anglerfish is among the most peculiar. Ferocious-looking sea monsters with oversized jaws and rows of sharp teeth, these fish have small extensions on their heads that resemble fishing rods with lighted lures, which hang down in front of their mouths. I watched a team of Russian scientists from the P. P. Shirshov Institute of Oceanology catch some of these creatures with fine mesh nets dropped to about 3,000 feet. To my amazement, the horrific anglerfish that came up from the depths was about the same size as my little finger.

On land, chemical olfactory signals play a key role in communicating information about territory and a readiness to mate. The potent smell that tends to get house cats neutered is a chemical signal that the female sprays on rugs and furniture as an invitation for males to mate with her. The ants that form a trail across the kitchen counter are premier biochemical experts, invading in columns after foragers have laid a chemical trial via little puffs of scent from chambers in their hindquarters.

As mate attraction signals, chemicals are used by most species of mammals. In some cases, chemical signals are essential for arousal. The male elephant dips its trunk into a potential female mate's urine to assess whether she is receptive, and then touches the tip of its trunk to an organ inside the roof of its mouth called the vomeronasal organ. If the female is in heat, a chemical in the urine will give the male an instant erection, the elephant version of Viagra. In fact, the female chemical stimulates production of nitric oxide, which is the same potent blood vessel dilator acted on by Viagra. If the female is not in estrus—the animal version of ovulation—no signal is received by the male elephant. Schooling fish will release chemical warnings when being attacked by a predator, but chemicals do not diffuse as well in water as they do in air, which is why land mammals use them to a much greater extent.

The ocean is instead a sensual universe of sound because sound waves travel more efficiently than light waves or chemicals through water. The aquatic environment has pressured its inhabitants to become vocal communicators and to develop abilities to hear and locate sounds from great distances. During a brief media fellowship at the Woods Hole Marine Biology Laboratory, I had the opportunity to examine the ear bones of a dolphin that had died after beaching itself. The bones looked nearly identical to those of the human ear except that they were several times larger. Their size reflects the dolphin's successful adaptation to the marine environment.

Dolphins and killer whales use both low-frequency sound waves and high-frequency sound bursts, many of which are out of our range of hearing. Most whale species communicate only with low-frequency sounds. Humpback whales are renowned for their haunting songs, which can travel long distances in the water, but the real long-distance communicators of the sea are finback and blue whales, which are the largest animals on earth, growing to 90 feet in length. The calls of finback and blue whales can be detected thousands of miles away. Like the songs of humpbacks, the vocalizations of finback and blue whales are probably designed to attract mates. Finbacks and blues normally communicate at depths of about 3,000 feet. There they can take advantage of a phenomenon known as the sound fixing and ranging (SOFAR) channel, a layer of the ocean that focuses the whales' low frequency sound waves and allows them to travel the greatest possible distances. The transmission of sound underwater is influenced by temperature, salinity, and pressure, and the conditions at 3,000 feet are such that sound waves become trapped and travel along the SOFAR layer for thousands of miles. The call of a blue whale is considered the loudest of any animal on earth, reaching levels of nearly 190 decibels, which is about as loud as a commercial jet taking off.

In the 1940s a young scientist named Donald Griffin discovered that bats navigate and locate their prey by bouncing high-frequency sound waves off the objects around them and analyzing the different rates of return of the sounds to their little brains. Griffin called this capability echolocation. Humans had developed a similar technology, which we call radar on land, and sonar, under water. At the time of Griffin's discovery, the navy was already using sonar, during

Sound Fixing and Ranging

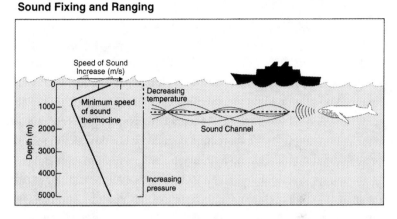

Whales make their long distance calls by exploiting a phenomenon known as Sound Fixing and Ranging (SOFAR). After having the SOFAR channel to themselves for millions of years, whales now compete with navy communications to get their messages across. How SOFAR works: The ocean is divided into horizontal layers of water of varying temperatures, salinity, and pressure. Colder temperatures slow the speed of sound while the greater pressures of lower depths increase the speed. Sound waves traveling through a region of rapid change in temperature from warmer to colder will bend downward as the speed of sound decreases. Greater pressures caused by depth refracts the sound waves back upward. This up-down-up-down bending of low-frequency sound waves allows the sound to travel hundreds of miles without the signal losing significant energy. (Source: National Oceanic and Atmospheric Administration.)

World War II, to detect submarines. After the war marine biologists, inspired by Griffin's discovery, borrowed the navy's sonar technology to explore communication in dolphins and whales.

In the mid-1950s, dolphin echolocation was only a theory. Marine Biologist Ken Norris's team figured out a way to blindfold a dolphin and conduct tests to demonstrate that dolphins have sonar. "He did a tremendous amount of work to understand how they do it," said Dr. Daniel P. Costa, a former student of Norris's and a professor of biology at the University of California at Santa Cruz.

Echolocation, which is used by most marine mammals, is not always used for communication. Marine mammals are better known for using echolocation, which sounds like repeated clicks to the human ear, to navigate and hunt. But in more recent years scientists have discovered that both bats and dolphins use their echolocation skills to converse. Marine mammal communication, especially in dolphins, is more complex than anyone had imagined.

On land, both sound and light are the big players in communication. Visual signals are far more important to land animals than sound because of the efficient movement of light waves through the air. Most terrestrial vertebrates have a well-developed repertoire of visual signals, including body language, markings, and flashes of color. Our human senses are limited to a middle range of sound and light frequencies. On the upper portion of the light spectrum is ultraviolet light. On the lower end is infrared light. Visible light falls between the two extremes. Birds can see ultraviolet light. Drab-looking feathers in the visible light range may appear like shimmering prisms to a little sparrow. My favorite snake, the fer-de-lance, a pit viper, can sense other animals' infrared radiation with organs called pits, which are covered with a temperature-sensitive membrane that detects body heat, similar to the way soldiers' night vision goggles work. Based on the amount of heat being generated, the snake can get a pretty good idea of the size of an animal that has come into its range.

Giraffes, elephants, and hippopotamuses are known to produce low-frequency vocal signals below our range of hearing to make long-distance calls. Probably many other species will be discovered to produce low-frequency sounds below our hearing threshold. With bats and dolphins chatting away at frequencies above our range of hearing, and large mammals holding conversations below it, it is obvious that a fair amount of animal communication is taking place beyond our sensory awareness. For at least half of the twentieth century, scientists tended to assume that if they couldn't see it, hear it, or smell it, then it must not exist, but they have since developed technologies that can hear sounds in the silence, see the invisible, and sniff out the tiniest whiffs of otherwise undetectable pheromones. What humans lack in sensory range and acuity, we make up for with our big, energy-guzzling brains—the neurological equivalents of SUVs.

No fancy technology, however, was needed to give research into animal communication its first really big breakthrough. Charles Darwin's era, the nineteenth century, marked the beginning of the formal study of animal communication. Scientists then focused on visual signals because they could observe them in the field and draw and reproduce them in books and journals for others to see. Darwin tended to discount the vocal signals of animals in his writings largely because he could not record sounds. With the invention of the tape recorder in the 1940s, however, researchers could finally go out into the field, make a recording of animal calls, and study them back in the lab. With the ability to listen to sounds in the lab, scientists could develop theories about their meanings, take the recordings back into the wild, and play them to the animals to observe how they responded. These "playback studies" launched the modern field of animal communication. Affordable lightweight digital video cameras are similarly revolutionizing the study of visual signals. The advance of digital technology has made research on both audio and visual signals incredibly productive. Digital audio and visual signals are much more readily broken down for analysis back at the lab, and digital files are easily transferred between scientists via the Internet.

I wish I had brought a digital tape recorder to record the rain forest's night sounds. Not only would I have had something soothing to listen to at home, I could have conducted a simple study of my own. I could have turned the tape recorder on, logged the time at 12:15 a.m., and marked my location with a handheld global positioning system unit. Back home I could have tried to identify the calls of the species I'd recorded and compared my recording with recordings of known species in the region. One of the world's largest collections of animal sounds can be found at Cornell University's Macaulay Library of Natural Sounds, which animal communication researchers use for this very purpose. (Anyone can log onto the sound library's Web site, http://birds.cornell.edu/lns/, and listen to animal vocalizations.)

My imaginary recording from the rain forest would be pretty busy, with hundreds of different creatures talking at once. With a machine called a sound spectrograph—another key breakthrough for communication scientists—I could break down the sounds into individual frequencies to analyze any vocal signal. The sound spectrograph, which may be more familiar as a machine used for studying speech and hearing disorders and music, consists of a tape recorder/playback unit, a device that scans the tape of your sounds, an electronic filter, and an electronic stylus—similar in concept to the ones used on old phonograph records. It transfers the analyzed sound information to paper. A spectrographic analysis looks similar to a printout from a heart monitor at a doctor's office or a lie detector, with lines on a sheet of paper that represent time, frequency, and amplitude (loudness) of the sound. Spectrographic analyses of the sounds of different species, and comparisons made between species, reveal some rather surprising similarities that I will discuss in detail in another chapter.

Katy Payne, a respected elephant behavior and communication expert from Cornell, demonstrated the power of sound spectrography in the early 1980s. She was visiting a zoo in Portland, Oregon, where two elephants had been separated from each other. Although the elephants couldn't see each other, they could still communicate. Payne said she happened to be standing near one of the elephants when she felt a strong vibration in her chest, similar to what one would experience standing near a pipe organ being played in a church. Payne recognized that the vibrations she felt were the result of low-frequency sound waves. She made recordings of the silent vibrations the next time she visited the elephants and demonstrated with the sound spectrograph that the elephants were communicating via low-frequency sounds. The findings launched a new field of study in elephant communication.

Payne was familiar with low-frequency sounds, which travel the longest distances of any frequency, from her studies in the 1970s of humpback whales and their long-distance calls. Her former husband, Roger Payne, is known for discovering that humpback whale vocalizations are actually songs. After she published her paper on elephant sounds, Payne collaborated with Joyce Pool and Cynthia Moss, who were already studying elephant behavior at the Amboseli National Park in southern Kenya. Together, they discovered much of what is known today about the meaning of elephant vocal signals.

One of the more novel inventions for studying animal communication comes from the laboratory of Peter Tyack at the Woods Hole Oceanographic Institute, in Massachusetts. Tyack and his students have invented an underwater microphone that can be attached to the heads of marine mammals with a suction cup. With this minimally invasive device, Tyack is collecting an enormous amount of data on vocal signals from marine mammals, including endangered right whales. The recordings of clicks and squeals made by these remarkable creatures sound like something a Hollywood special effects team might conjure for a space alien, and they are expected to yield important new insights into the behavior of right whales, which might enable scientists to reduce the number of collisions that occur every year between these whales and boats. Another of Tyack's inventions, which lights up when an animal is vocalizing, can be attached to the heads of dolphins for studies of dolphin interactions in captivity.

Animals do not need any special equipment to be expert sound and light engineers. They are natural physicists that exploit light and sound in sophisticated ways. To keep up, animal communication experts must learn the science of light and sound, how they are propagated and received, before they go out into the field to begin recording and filming their subjects.

Scientists have discovered that animals are experts at exploiting weather conditions and the physical conditions of their environments so that they are heard or not heard, and seen or not seen. The species living here in the rain forest of Peru must engineer their calls to accommodate all of the obstacles, such as leaf cover, that can deflect and degrade the sounds intended for a potential receiver. Overall, short, loud bursts of sound tend to be more effective than longer calls at cutting through the dense foliage.

There is no natural environment on earth noisier than a virgin rain forest. Every species here has developed clever or remarkably sophisticated strategies to ensure that its voice is heard. The noise creates a real challenge for the smaller residents, such as male tree crickets, which need to get the attention of females, often from a relatively long distance. Some species of crickets maximize the volume of their stridulating calls by chewing a hole in the middle of a leaf to create a sound baffle, in the same manner humans build a stereo speaker. The leaf functions as a speaker cabinet, with the cricket in the center acting as the speaker.

A species of tree frog in Borneo has a unique approach to getting its mating call heard over the din. *Metaphrenella sudana,* which is only an inch long, has learned to exploit the sound properties of a water-filled hole in a tree in the same way that a person uses resonance in the shower to sing like a star. The frog searches for a suitable hole and then partially submerges itself in the water. Its forte is the ability to adjust the frequency of its call to the size of the hole and play the tree like a musical instrument. As it sits in the hole, it begins vocalizing at different frequencies—lo lo lo, la la la, le le le— until it hits the one note that makes the hole and tree resonate.

The time of day affects how sound travels in any environment, and this fact is not lost on animals and insects. Early morning and late evening produce conditions that allow sound to travel greater distances than during the middle parts of the day. Sound travels best at night, which is why the rain forest is so wonderfully noisy between dusk and dawn. For species that sleep at night, dusk and dawn are their windows of opportunity to get the best resonance and distance out of a signal. This is why animals, especially birds, tend be more active and noisy in the early morning and late evening. The British call the phenomenon of birds singing in the early morning the dawn chorus. Because of the superior sound conditions, dusk and dawn are the times to conduct the serious business of attracting mates and defending territory. For predators, it is the best time to eavesdrop on conversations and track down their chatty prey.

Another way animals and insects ensure that their calls connect with the intended receivers is by developing their own specialized frequencies, which are determined primarily by the size of their bodies. Recently, a scientist visiting near this region of the rain forest made an audiotape of a little of the night's music. When he took the tape back to his lab and analyzed it, he discovered that this seemingly chaotic banquet of sound is actually highly ordered. Each animal and insect is tuned to and calling on its own species-specific frequency, in the same way that radio stations use different signals so that many stations can be on the air at the same time. You might say that the crickets are playing on WCHIRP FM 102, while the tree frogs are at WCROAK FM 97.1. Mosquitoes, which really bug me, buzz at a perfect high C.

Bernard Krause, a professor at the University of Oregon in Eugene, has found that in older tropical rain forests some species, such as the Asian paradise flycatcher, have become so specialized that their voices occupy several niches of the sound spectrum at the same time, "thus laying territorial claim to several audio channels." His recordings from undisturbed rain forests around the world demonstrate a remarkable stability in the combined voices of the residents from year to year. The stability of the ambient sound gives each region a unique sound signature, or fingerprint.

"Over a number of years we would return to the same sites," Krause says, "only to find, when the recordings were analyzed, that each place showed incredible bioacoustic consistency, much like we would expect to find from fingerprint matching. The bird, mammal, and frog vocalizations we recorded all seemed to fit neatly into their respective niches. And the bioacoustic niches from the same locations all remained the same given the time of year, day, and weather patterns. Having just begun to work in Indonesian rain forests, early analysis indicates similar results from each of the biomes we have visited and recorded."

Krause compares the sounds of the nighttime jungle to the music of a symphony orchestra. Different instruments are tuned and played at specially timed intervals to avoid drowning each other out. Animals do the same thing by finding their niche frequencies and taking advantage of the pauses that others make during their calls. Even though the sounds seem chaotic to the uninitiated human ear, all of the vocalizations are actually highly choreographed. Of course no rain-forest conductor coordinates the thousands of voices booming in the night. The orchestration of the voices of thousands of animals and insects calling together developed over time, as each species fine-tuned its signals and reception for maximum efficiency.

According to Krause, the Jivaro and other tribes of the Amazon Basin have known this information all along. These indigenous people are expert at identifying the sounds of individual animals as well as subtle differences between various mini-habitats in a region. Moving as little as 30 feet in one direction in an old-growth forest will reveal an entirely different ambient sound signature. The local tribes are able to travel at night in total darkness and identify their location solely by the sound. Krause believes that ancient humans would have possessed the same intimate knowledge and familiarity with their environment. His research suggests that the roots of musical composition can be found in the animal orchestras of the old-growth rain forests.

Comparisons of the vocalizations of songbirds to music are not meant as a metaphor. Birdsong is music. It is supposed to sound pretty. The emotion that birdsong evokes in a human listener is not far from the physiological response it is intended to evoke in the female receiver. Of course, female receivers are evaluating the songs for much more than just their nice sound. They are judging the quality of the songs and the repertoire, which scientists believe reveals information about the quality of the singer's genes. (Some female human concertgoers also evaluate rock singers' jeans and determine their own willingness to mate based on the quality of the songs and repertoire.)

Why did animal communication first arise? The prevailing theory, credited to the classical ethologist Niko Tinbergen, is that animal communication evolved as a more economical substitute for physical violence. If animals can growl and posture and bluff instead of hurting each other and still get what they want, then why get hurt? Injuries from fighting make it harder for an animal to conduct its business of mating, defending territory, and finding dinner. Life in the wild is risky enough as it is. Violence is very expensive in nature's economy, and this is why animals rarely come to blows. Any successful strategy that avoids violence will be favored through natural selection. The injured hotheads will die off while the successful bluffers mate and pass along their genes.

Tinbergen and Konrad Lorenz developed the concept of "intention movements" to explain the evolution of communication, suggesting that the signals an animal uses to communicate in a conflict arose from the same physical movements and vocal sounds that it would normally use during an actual fight. For example, many mammals reflexively retract their lips when fighting to protect them from being bitten off by an opponent, and they retract their lips when biting another animal. That reflex has evolved into baring one's teeth at a potential opponent. (The human snarl curls the lip to reveal a canine tooth, which was larger in our ancient ancestors.)

During a conflict, the natural fear and aggression an animal feels will cause its hair, assuming it's a hairy beast, to stand on end. This autonomic, or involuntary, response to fear and aggression, called piloerection, has become another bluffing signal in a conflict—it makes the animal look bigger. The idea is that if an animal appears as if it intends to fight by assuming the various postures and vocalizations associated with fighting, maybe the opponent will back down.

I will present all of these modes of communication in greater detail in later chapters. In the meantime, as I prepare to slip away from camp, the sounds are particularly intoxicating. But it is February and the mosquitoes at this time are abundant. Outside the protective netting of my hammock, these miniature vampires quickly engulf my head like a thick cloud, filling my ears with their high-pitched, annoying *"bzzzzzr, bzzzzzr."* Rain-forest veterans warned me about them, but I had no idea there would be so many at all hours of the day. Mosquitoes have short lives, so I suppose they must make the best of it, but I feel as if I've been tricked. In all of the Tarzan movies I saw as a kid, never once did the king of the jungle slap a blood-engorged mosquito off his neck and curse. I have lathered so much Deet insecticide onto my skin that I can smell the stuff on my breath. Even so, my ears, which protrude from my head like satellite dishes, are already burning and itching. With a shudder and a repressed desire to slap the sides of my head, I sling a light pack over my shoulder, tuck my pant legs into my boots, and slip out into the night for the party. A mere 20 feet from the edge of the camp's platform, the clearing yields to the forest. I can barely see what appears to be a dark hole that might be the trail leading to the canopy walkway about half a mile away.

My heart pounds faster as I move into the near total darkness of the surrounding jungle. Feeling like Alice stepping through the looking glass, I stumble blindly in the dark along a muddy trail until I am at a safe distance from camp. Finally I turn on my rubber-coated flashlight to illuminate the narrow path, no more than a couple of feet wide. A sense of marvel and the distinct tingle of fear sweep through me as I head deeper into the jungle. The scientists did not caution us to stick close to camp for nothing. Fer-de-lance could be anywhere, and I have to be careful not to brush against branches or step on something long and slender that is waiting for small prey to come along. After last night's encounter with the viper I admit developing something of a preoccupation with this cousin of the bushmaster and rattlesnake. Brown, sometimes gray, with light stripes and diamond markings similar to those of rattlesnakes, the fer-de-lance has a distinctive yellow throat and jaw, which communicate a simple message: "I'm poisonous. Don't try to eat me."

In nature, colors are very simple codes with specific, universal meanings. Yellow and red (and sometimes orange) are the animal kingdom's favored colors for advertising that a particular snake, insect, frog, bird, or other critter is venomous or at least not very tasty. I especially do not want my flashlight to pick out the yellow throat and open mouth of a fer-de-lance that has raised itself up from the ground in a taut S-shape. That would mean I—a large primate and potential predator—have frightened the poor fellow and made it ready to defend itself. Given that communication arose as a substitute for violence, the snake will try to warn me before using up its valuable venom—if I see it. As a city dweller, I am at a distinct disadvantage in the jungle at night. Most of the animals, including the tree frogs and crickets, can see me, hear me, smell me, or feel the vibrations of my feet on the path long before I am aware of them.

If I possessed infrared sensors, I would be able to see the many creatures that are certainly watching me. I decide to turn the flashlight off and stand in the darkness for a few minutes to soak up the sounds and get a feeling for how my own inferior senses work in this unfamiliar environment. My sense of touch tells me that the mosquitoes are continuing to attack my face, but my eyes can just make out the shapes of trees and undergrowth although not much else because the thick canopy blocks the moonlight. Sound is about the only thing my senses can pick up, and it is overwhelming since I cannot distinguish the different animal sounds with anything near the skill of the Jivaro.

After ten long minutes in the darkness I flick the flashlight back on and continue my illicit walk down the narrow trail. The flashlight restores my human vision and confidence. The moment I light up the trail, hundreds of flying insects are attracted to the jiggling beam. Quite a few theories have been developed about why insects, especially moths, are attracted to light. Some scientists say it's because they use the moon to navigate and confuse the brighter rays of a porch light or flashlight with their lunar guide. Curiously, the same phenomenon occurs in the ocean. During a night scuba dive, tiny copepods—basically the bugs of the sea—swarm around a flashlight just like these insects. Copepods migrate toward the surface with the first light of day, and their sensors confuse the flashlight with the sun. Here on the trail, my flashlight beam creates an opportunity for bats that quickly zero in on the halo of bugs. I fancy I can almost hear their high-pitched sonar as they ping the bugs and dart in for an easy meal.

The calls of several douroucouli, a very vocal nocturnal monkey, pierce unexpectedly through the upper canopy. The douroucouli are tiny primates with big eyes, specialized for night vision, that have earned them the nickname of owl monkey, the only species of monkey known to be active at night. Their call sounds like a high-pitched "wook wook" and carries a distinctive sense of urgency— the douroucouli's alarm call. Most species of primates regard their human cousins as predators, and these guys are definitely alerting each other to my presence on the trail. I don't know whether they heard me, smelled me, or saw the flashlight first. Either way I feel guilty for upsetting them. Local people hunt them for food and sell their fur. The monkeys are also captured and sold as pets and for use by pharmaceutical companies. The group of monkeys I have alarmed is most likely a family. Males and females mate for life and generally have two to five youngsters at any given time sponging off Mom and Dad at home. Dads carry the little ones on their backs and take responsibility for most of the childcare. These monkeys are extremely territorial and mark their boundaries by means of a gland at the base of the tail that excretes an oily brown liquid. Owl monkeys make about 50 different calls. They squeak, hiss, bark, hoot, and meow, and all the calls have quite specific meanings, including alerting each other to a food cache, rallying together, and bonding.

Alarm calls are especially common in primates. Captive monkeys used for experiments at medical labs quickly adapt their alarm calls to veterinarians and technicians who perform invasive procedures. The monkeys can make a disturbing fuss on the days that blood must be drawn or when one of them is removed and taken to surgery. Whenever a stranger shows up in their midst they go absolutely bonkers with angst because they have no idea what to expect and tend to associate unfamiliar humans with pain. Their cries of fear are unmistakably anguished and can provoke sorrowful emotions in empathetic lab workers and students unaccustomed to such calls. Of all the sounds in the animal kingdom, it is the cries of primates that I find the most disturbing. But closer to home, as the West Nile virus spread from crows to squirrels, people in Illinois began reporting sounds like a baby crying in pain. It turned out to be squirrels that were infected with the virus and experiencing neurological degeneration.

I move on toward the walkway and startle other animals as my boots squish into the muddy trail. Crickets, toads, and tree frogs abruptly cease their calls as they sense my approach, but resume their business of courtship and territory defense as soon as I pass by. Tree frogs make up a large section of tonight's orchestra. I amuse myself with a little game of stop and go—silencing the crickets and frogs, then waiting for them to start in again. Funny how one can sit in an apartment in a city surrounded by marvels of man-made communication technology—telephones, faxes, cable television, music CDs, and high-speed Internet—and be bored to tears. It is much more fun to interact with the frogs.

After about a half-hour walk, I reach my destination at the end of the trail. Stretching overhead is a maze of steel cables, ladders, and webbed rope bridges that ascend to the top of the canopy. I climb a sturdy ladder through the first layers of growth to a small platform that leads to a series of rope bridges and more ladders. A green tree snake slithers over a limb and out of sight, startled by my sudden appearance. I am reminded not to lean against branches or tree trunks and to make sure I can see where I'm placing my hands. The mosquitoes mercifully thin out with the altitude, and moonlight begins to break through the thick foliage as I approach the platform at the summit. The top consists of a bridge, constructed of rope webbing for sides and an aluminum bottom, which spans about 30 feet between two trees.

At last, the unbroken rain forest stretches before me to the horizon in every direction. Seeing it is a profound and moving experience. The humid, floral-scented air vibrates wildly with the primal chorus of exotic tropical screech owls, lyre-tailed nightjars, short-nosed tree rats, kinkajou, douroucouli, cicadas, and probably a few creatures still unknown to man. The night is thrillingly alive and stares back with glowing orange eyes from the limbs of giant mahogany trees, acacias, and thorny palms. Here, standing at the top of this stairway to heaven under a brilliant three-quarter moon, one can hear the unbridled voices of nature and glimpse a vision of a primordial planet from some distant epoch teeming with strange and wondrous varieties of life. Hearing so many different voices is like stepping out of New York City's Penn Station for the first time and being bombarded with shouts in Spanish, Hindi, and Vietnamese, honking taxis, strange smells, sexually enticing billboards, and blazing neon signs. The seemingly chaotic sounds of the jungle remind me of the biblical story of the Tower of Babel.

A metaphor in discussions about the origin of language, this story in Genesis recounts how people have gathered from the wilderness to create a new city after the great flood that destroyed civilization. They build a tower that reaches to the heavens to "make a name for themselves" so they will never be dispersed again. God does not like their plan, so he destroys the city, scatters everyone to the far ends of the earth, and makes them speak different languages.

The opening sentence of this biblical legend refers to the period before the people came from the wilderness and started building the city of Babel and their own stairway to heaven. As I stood on the canopy walkway surrounded by the voices of nature, the sentence began to make sense in a rather odd way:

"Now the whole earth had one language and few words."

The ancient Greeks referred to animal vocal sounds as the natural language. Might this wonderful and tightly choreographed orchestra of voices in the rain forest represent a single language that all the species understand? Perhaps monkeys and birds understand each other. If this is true, how could it be? Clearly, different species make distinctly different vocalizations, but they also share the same five senses to a large extent and have similar needs. Those that live together in the same environments had to deal with the same physical properties of light and sound when developing their signals over the ages. Similar environmental conditions would tend to cause vocal signals to bear at least some similarities in their structure, and it turns out from comparisons of sound spectrographs of different species' calls that this is true. For reasons you will see as we move along, the forces of natural and sexual selection have driven the communication of widely divergent species down similar paths.

Consider human language. No matter what region of the world in which we might have been raised, and no matter what types of words we use to describe the things around us—German, Russian, English, or Japanese—we all speak a human language and follow the same rules of grammar and syntax. We might not understand the words that another person is speaking, but we can usually determine whether that person is hostile or friendly or upset, and by using gestures and various emotional expressions, we can often get the essence of what that person wants. Animals of different species sound different as well. Dogs bark, birds chirp, and horses whinny, but they are astute at understanding each other. Animal communication is not as complex as human language because there is less information to be conveyed in an animal's daily life. But humans and animals alike, regardless of race or species, talk about the same things every day—that is, sex, real estate, who's boss, and what's for dinner.

The whole earth does have one language with few words, and all species, including humans, continue to use it every day. It is a natural language that stems from the evolutionary roots shared by each of the 10 million species that inhabit the planet. We may see ourselves as quite separate from the animal world, but as we move along I hope to show that we still have a seat in the orchestra.

Two

OF RAVENS AND ROBOTS

W HAT KIND OF animal communication book would this be without at least once invoking Dr. Dolittle, who embodies our fascination with the yips, chirps, screeches, and hoots of animal talk and our desire to understand what it is they are saying? *The Voyages of Doctor Dolittle,* first published in 1922, is the story of a medical doctor whose love of animals destroys his practice but launches him into a series of great adventures. Dolittle had learned to speak animal language fluently with the help of Polynesia, his acerbic parrot. The turning point in his life came when a sparrow flew to his English cottage from Africa to plead for the doctor's help with an outbreak of disease on Spidermonkey Island.

On more than one occasion in the course of researching this book, I too felt like pleading for Dr. Dolittle's help, but with all due respect to the ideal he represents, such remarkable progress has been made in animal communication that I daresay his services are no longer required. A new generation of scientists has emerged, including experts from evolutionary biology, behavioral ecology, genetics, linguistics, psychology, philosophy, and physics, all contributing to an increasingly rich appreciation for the intelligence and emotions that lie behind those animal eyes.

In 1920, when Dr. Dolittle first captured the hearts of people worldwide, the field of animal behavior was dominated by comparative psychologists, who viewed animals as unthinking creatures, incapable of emotion and driven purely by instinct. The popularity of Dr. Dolittle contrasted with the scientific attitudes of that time and revealed an immense chasm between laypeople and scientists regarding their perceptions of animals. That chasm persists today, but it is growing narrower. To understand the new science of animal communication and its current controversies, it is important to peer behind the curtain and take a look at what has influenced scientists' attitudes over the years. Why would so many of them reject what every pet owner knows to be true—that animals are thinking and even compassionate creatures?

The traditional view that animals are biological automatons can be traced back 400 years, with the roots of today's scientific dogma, though changing, taking hold in the late 1800s. Yet although scientists can be accused of believing that animals are furry robots, some members of the public can be taken to task for believing that animals are furry humans. Animals are neither. They are unique species with their own mind-sets. When studied objectively with modern scientific tools, they reveal many similarities to us humans in behavior, physiology, and nonverbal modes of communication. Our differences can be measured in degrees rather than in kind.

Let's begin with a story that actually falls between the two extremes of attitudes held by scientists and the public. Bernd Heinrich, a scientist at the University of Vermont who has studied and raised ravens for more than 20 years, has a tremendous appreciation for animal cognition. He could also be described as an animal lover. In one of his books, *Ravens in Winter*, he recounts the story of a woman's encounter with a raven and a mountain lion. The woman lives next to a canyon in a rural area outside Boulder, Colorado. As she was working one day in her garden, she heard a raven squawking somewhere out of sight. Many people consider ravens a nuisance, even though they are perhaps the most intelligent of birds. She paid it no attention and kept gardening. The raven continued calling, and after about half an hour the woman was growing increasingly annoyed by its raspy voice. The raven's calls had become louder and it had begun circling overhead, flying nearer to her backyard, which abutted the canyon. Finally, the noisy raven flew into the yard directly above her and landed on a large rock, where it perched and gazed at her with its black eyes. Irritated, she glanced at her unwelcome guest and noticed a slight movement in the bushes near the rock. There stood a large mountain lion, only about 20 feet away, crouching in a predatory position, seemingly prepared to pounce on her small frame. The terrified woman began backing away and calling for her much larger husband to come to her aid. (In the field of animal communication, this is known as a distress call. Animals also call for help from mates or group members when facing a predator.) Her husband came running out of the house and was able to scare away the mountain lion. Perhaps from its perspective, the moment of surprise had been foiled and it would have had to expend too much energy competing with another large mammal for its dinner.

The story made the local news, and the woman related how she believed that the raven had made the terrible fuss to warn her of the mountain lion's attack. If the raven had not been vocalizing so loudly and persistently, she would not have paid attention, and she thanked God for sending the raven to alert her to the danger. The event was reported by the local media as a heartwarming story about the raven saving the woman's life.

People seem to love these kinds of stories, and rightly so. They help build a bridge across the gap in perceptions and create a bond between humans and nature. In 1996 a three-year-old boy fell into a gorilla enclosure, injuring his head, at a zoo in Brookfield, Illinois. A seven-year-old female mountain gorilla, named Binti-Jua, gingerly lifted the boy into her arms and placed him near a door of the enclosure so that zoo officials could take him and administer aid. There is no doubt that Binti-Jua felt compassion for the young primate of a closely related species and intended to help rescue him. At the time, her own offspring was clinging to her back. The story made international news.

But Heinrich has a different take on the raven story, based on his years of experience and research. Heinrich said, "A raven warning a human about danger simply doesn't make sense in nature." Ravens are known to guide predators to prey. The more likely scenario, according to Heinrich, is that the raven had sighted the woman in the yard from its bird's-eye view, saw the mountain lion at some reasonably close distance to the woman's yard, and decided to play matchmaker. Ravens and crows are quite clever problem solvers, animal Einsteins. Attracting the mountain lion's attention and directing it to the yard had consumed at least half an hour of the raven's time and energy. Since economics plays a central role in the behavior of wild animals, the raven must have expected a fat return on its investment. The dividend would have been a carcass to scavenge after the mountain lion had filled its belly. One of the reasons ravens gained a bad reputation is that our forebears noticed that they do not discriminate between human and animal carcasses.

Without the kind of information Heinrich has collected over 20 years of research and personal observations, people assumed that the raven was giving a warning call to save a human, just like something out of an old fairy tale. When we interpret natural behaviors through personal belief systems, however, we risk missing the truth. More important, we miss the opportunity to ask questions, such as, how does a mountain lion arrive at the decision to interrupt its business to follow a raven, and how did the raven learn that it could influence the predator's behavior? Had this particular raven done this before with other predators and prey? Did the raven's calls convey information that the mountain lion understood as "Follow me—got a fresh meal for you"? Or did the raven's excited chatter simply get the cougar's attention, and because of some past experience similar to classical conditioning, the mountain lion made an association between the raven and a prey item? The raven and the mountain lion certainly appeared to be communicating to some extent with each other, and that is worthy of exploring scientifically.

These are the types of questions that scientists are asking about many different species. Almost monthly, new evidence emerges to show that animals engage in a variety of complex behavior that requires thinking and decision making. I believe the most remarkable of these studies involves Betty, a New Caledonian crow and a hot star of animal research whose mental prowess has been captured on video. Betty is the traditional skeptic's worst nightmare and has taken animal cognition to a new level.

Betty resides at the Behavioral Ecology Research Group laboratory at Oxford University in England. A team of scientists led by Professor Alex Kacelnik who studied tool use in animals had arranged an experiment to find out whether Betty and an older bully of a crow named Abel could select the proper tool for a job if given a choice of tools. New Caledonian crows are avid tool users in the wild and frequently use leaves, twigs, and feathers to catch prey. For the experiment, Betty and Abel had to choose between a straight piece of wire or a hooked wire to snag the handle of a little bucket containing a piece of meat at the bottom of a tube. Both Betty and Abel quickly determined that the hooked wire was the best tool for the job—kid stuff, actually, for crows. But during the experiment, Abel, being bigger and dominant, stole Betty's hook. Without hesitation, Betty picked up the remaining straight wire with her beak, wedged the tip in a crack on the laboratory table, and bent it with her beak to form a hook exactly like the one Abel had appropriated. Betty then proceeded to use her hook to snag the bucket in the tube and retrieve the meat.

Kacelnik says that Betty's spontaneous performance using an object not found in nature is one of the most stunning examples of animal ingenuity and problem solving ever observed—greater than the toolmaking skills of wild chimpanzees, which use straw or twigs to dip into ant and termite nests for a meal. Chimpanzees have also been observed crushing hard palm nuts with rocks. But no primate other than a human, or any other animal, except Betty, is known to have taken a man-made object and fashioned it with no prior experience into a tool for a specific purpose. Kacelnik said he did not bother to find out whether Abel could repeat Betty's performance because old dominant crows have a different mind-set—they simply wait for subordinates to do the work and then take their food.

Betty demonstrated that she fully understood the purpose of the tool and was able to create a mental image of the stolen hooked wire to fashion a new hook. Marc Bekoff, a scientist at the University of Colorado in Boulder, who studies the behavior of wild coyotes and who has pioneered research into animal play behavior and emotions, described the experiment as "perhaps one of the most significant studies in animal thinking of any species."

The story of the raven and the mountain lion, while anecdotal, supports the idea that some species of wild animals can develop a goal and execute plans to achieve it. One of the more contentious, long-standing debates among scientists is whether communication is intentional. Betty showed clear intent in her behavior, which suggests that the nefarious raven also knew what it was doing by calling repeatedly to the mountain lion. One anecdote does not make a case, but there other examples of partnerships between different species that reveal goal-oriented behavior and willful communication. The best known of these partnerships involves a species of bird that leads animals and people to beehives so it can share the honey. The honeyguide, also known as a black-throated indicator, resides in African and South American rain forests. A relative of woodpeckers and toucans, it is the Winnie the Pooh of birds and, true to its name, has a great fondness for honey and bee larvae.

The diminutive, plain-colored honeyguide, which is only the size of a small lark, is unable by itself to break apart a bees' nest to get the spoils.

One of the earliest written accounts of its behavior came from a British traveler in South Africa in 1777. How the behavior originated is unknown, but at some point in the species' past, an especially clever black-throated indicator must have observed that Pygmies and an animal known as the honey badger, or ratel, shared its culinary tastes and possessed the ability to open a hive. Perhaps that bird, with its fondness for honey, spent a fair amount of time lurking around bees' nests and happened to be present one day when a badger or Pygmy was breaking open a nest. Animals quickly learn associations between a behavior and a prized food. A successful new behavior, struck upon either by accident or ingenuity, usually spreads quickly through a population.

A classic example of this occurred in the 1930s with blue tit birds in the British Isles. In those days, milk was delivered to people's homes early in the morning in bottles capped with cardboard. The milk was not homogenized, so the cream would separate and rise to the top. According to James Gould, a behavioral ecologist at Princeton University in New Jersey, blue tits naturally forage by peeling away bark to get at insects. In captivity they will do the same thing to wallpaper. The behavior is innate. One early morning, a blue tit most likely lit on top of a milk bottle by chance and reflexively peeled back the cardboard cap of the bottle. Striking the animal equivalent of the lottery, it skimmed the cream from the top of the bottle. When one bird sees another bird helping itself to a meal, it will barge in and join the feast, attracting other birds. (A similar phenomenon occurs in my newsroom when someone is having a birthday. The moment reporters realize that others are having some cake, they flock to the party.) People began reporting to the milk company that their bottles were being opened and the cream taken. The reports spread steadily from the first location outward to the entire region, and then across Britain, according to Gould. Thousands of blue tits in a relatively short period of time were in on the action. The milk company countered by putting foil caps on the bottles, but this did not deter the birds. It took the milk companies in England nearly ten years to design a carton that the birds could not raid.

The blue tit's behavior is not as sophisticated as the honeyguide's, but it is a good example of how a behavior can get started. The honeyguide's strategy is to locate a hive, fly off to find a Pygmy or a honey badger, and lead one or the other back to the hive to break it open. To accomplish this, the honeyguide communicates directly with its potential business partner by chirping loudly and fluttering until it has its partner's undivided attention. On the way to the hive, the bird is careful not to fly ahead too quickly. The Pygmies, being considerate associates, always leave enough of the honey, beeswax, and bee larvae behind for the bird's dinner. The badger, on the other hand, eats its fill, but there seem to be enough leftovers to keep that partnership alive too. What is remarkable about the honeyguide is its successful communication with two different species—humans and ratels. One might say the honeyguide, the ratel, and the humans are communicating with the same animal language, however simple it might be. They certainly demonstrate an understanding of each other's wishes and cooperate to achieve a common goal.

These examples do not necessarily indicate that animals think like humans, but there is no doubt that the animals are using their brains. Nonetheless, had Betty demonstrated her feat to scientists only 15 years ago, many would have scoffed and called it a fluke, even with the video documentation. The argument would have been: "Even though the crow appears to be thinking, one must prove conclusively that it was not simply acting on an instinct already programmed into its genes. The behavior may be similar to one in which the animal engages in its own habitat." Although this is true of the blue tit and the milk bottles—peeling back material to get at food is what it does naturally—Betty's case is quite different. Under the old way of thinking, scientists would have spent time and resources to prove that Betty can think rather than designing experiments to see what else she can do.

The French philosopher and mathematician René Descartes is often blamed for the attitude that animals are thoughtless, soulless automatons. Descartes's famous phrase, *Cogito, ergo sum,* "I think, therefore I am," is where the trouble began. The attitudes that have pervaded animal behavior and communication research for more than 100 years originated from the classic interpretation of Descartes's view of man and animals. Descartes provides the basis for what is known today as the theory of mind, which refers to the ability of one individual to attribute mental states to another individual. For example, as young children develop, they reach a stage at around age three when they distinguish between themselves as individuals and others and become more sensitive to others around them. When they recognize that others have their own thoughts and feelings, they achieve a "theory of mind." One of the contentious debates in the field of animal behavior is whether animals possess a theory of mind.

According to Descartes, human beings are *res cognitans,* or "thinking substance." Everything else is *res extensa,* "extended substance," not capable of consciousness or thought. In 1637, Descartes published the lengthy *Discours de la Methode,* a revolutionary text in its establishment of the notion of individual identity, particularly notable given the political climate of the day and the disparity between the rich and poor classes. With this treatise, Descartes also broke from the broadly accepted philosophical doctrine known as the Great Chain of Being, which held that everything in the universe was divine and linked together in a continuum that progressed from inorganic matter, such as rocks and minerals, to plants, animals, man, demons, angels, and ultimately God. The Great Chain of Being created a hierarchy of increasing complexity and was the model all educated people used to describe themselves and their world from the period of the ancient Greeks until the seventeenth century.

Descartes's daring notion set humans apart from everything in nature because humans possess a mind and thus a soul and, in his view, nothing else does. From Descartes's perspective, the mind's awareness is solely responsible for creating the world around it. Without such judgment or awareness, humans would be machines. Descartes and most other educated people in seventeenth-century France had developed a fascination with the emerging technology of machines and with the prospect that automatons could be created to do the labor of humans and animals. The only way to determine whether another person has a mind is to speak to him. If he responds, then one can assume he has a mind.

Descartes's revolutionary ideas changed the way human society viewed its place in nature. People were no longer part of the continuum, or Great Chain of Being. Possession of a mind and thus a soul made humans more like God. Animals lost their divinity and became machines. Descartes wrote further that humans have both voluntary and involuntary movements. Breathing is an involuntary movement, performed without thinking. Voluntary movements, which Descartes assumed included language, are under the control of the mind, or will. He regarded all actions of animals as involuntary movements. Since animals do not speak human language, Descartes assumed they had no will.

The medical profession in Descartes's time, as well as the general society, was eager to embrace the philosopher's way of thinking. At that time, physicians, at least in France, were prohibited from performing vivisection experiments (live dissections) on animals because the Great Chain of Being held that all creatures are divine and part of each other. The prohibition had made it impossible for physicians to study anything but dead animals and human cadavers, assuming they could get their hands on one. Physicians wanted to see how live organs performed and learn how bodies function as a system. Dead animals and the occasional corpse provided information only about anatomy. If Descartes's ideas were accepted, and animals were considered unthinking, lacking souls, and thus unfeeling, then experiments on live animals could be justified. Gruesome, cruel experiments were sanctioned and performed. Animals that were writhing and vocalizing expressions of pain during experiments were not considered capable of actually experiencing pain. Vocal expressions of any type were viewed as the automatic responses of the machine. Incredibly, quite a few scientists still have this inhumane belief.

Descartes's views initiated a sharp division between humans and animals. The idea that humans were separate from nature became ingrained in European society and eventually in the United States. More than 200 years after Descartes, Charles Darwin came along to shake things up. After Darwin published *The Origin of Species* in 1859 and the scientists of his day came generally to accept the theory of evolution, he published another treatise, *The Descent of Man, and Selection in Relation to Sex,* in 1871. In *The Descent* Darwin presented evidence for the theory that humans had descended from apes. Scientists had been willing to accept Darwin's ideas about the origins of species, in part because they did not conflict with the Cartesian perspective that humans and animals were separate, but the idea that people were descended from apes was too much. Darwin wrote:

Nevertheless the difference in mind between man and the higher animals, great as it is, certainly is one of degree and not of kind. We have seen that the senses and intuitions, the various emotions and faculties, such as love, memory, attention, curiosity, imitation, reason, etc., of which man boasts, may be found in an incipient, or even sometimes in a well-developed condition, in the lower animals.

The scientific community became sharply divided and confused. Man as a descendant of apes presented a dilemma for scientists who accepted the theory of evolution but who also believed in Descartes's view of animals as unthinking biological machines. It also created problems for scientists who believed in God but who saw the logic of Darwin's argument. The troubling question arose: If apes have no mind, how did humans come to possess a mind and language? How is it possible to acquire a mind from nature where it does not exist? Descartes believed that God provided man with his soul and language in accordance with the biblical story of creation. Still hotly debated, this question lies at the center of the controversy about whether apes, or any other animals, possess a theory of mind.

Scientists of Darwin's era became more polarized and continued the debate while Darwin moved on to another interest—the origins and expressions of emotions. In 1872, during the latter part of his career, Darwin published what is arguably the first book on animal communication, titled *The Expression of the Emotions in Man and Animals,* which led to another brief respite from the Cartesian perspective. Most important about this work is that it detailed the facial expressions and body language of various animals and, for the first time, associated specific facial expressions and body movements with emotional and motivational states. These observations provided the raw materials for later scientists, particularly the classical ethologists Niko Tinbergen and Konrad Lorenz, to develop their ideas about the origins of communication and the evolution of signals. To some degree, Darwin demonstrated that emotions are universal in humans and animals—at least in mammals. *The Expression of the Emotions* illustrates in great detail how humans and animals express emotions in a similar manner.

If humans and animals express emotions similarly, it should be possible to recognize each other's emotional and motivational states. My mom's dog, Taco, knows when she is feeling ill. When her ankles swell with fluid because of her congestive heart disease,

Taco will gently lick them. The ability to recognize emotional states among wild animals would be strongly favored by natural selection because it would allow animals to avoid conflicts and form bonds; it would be favored by sexual selection in attracting mates. While Descartes insisted that two people must be able to use human language for each to recognize that the other has a mind, language is not necessary for different species to recognize one another's pain, grief, joy, hostility, or appeasement. This emotional recognition across species boundaries, to which we will return, provides the foundation for a universal animal language.

Darwin's *Expression* sold 9,000 copies in the first four months, a best-seller. But it sparked a sensational interest in animal emotions that likely exceeded anything Darwin could have anticipated and gave rise to an unfortunate era of wild anthropomorphism. George Romanes, a popular and avid naturalist, followed Darwin's example and began seeking stories of interesting, unusual animal behavior from numerous sources throughout Europe. Romanes took pains to ensure, as much as possible, that the sources were reliable. In 1888 he published *Animal Intelligence,* which was flush with great anecdotes, good intentions, and more than a few misguided conclusions. One example from Romanes:

> But that some species of ants display marked signs of what we may call sympathy even towards healthy companions in distress, is proved by the following observation of Mr. Belt. He writes: "One day, watching a small column of these ants (*Eciton hamata*), I placed a little stone on one of them to secure it. The next that approached, as soon as it discovered its situation, ran backwards in an agitated manner, and soon communicated the intelligence to the others. They rushed to the rescue . . . Another time I found a very few of them passing along at intervals. I confined one of these under a piece of clay at a little distance from the line, with his head projecting.

Several ants passed it, but at least one discovered it and tried to pull it out, but could not. It immediately set off at a great rate, and I thought it had deserted its comrade, but it had only gone for assistance, for in a short time about a dozen ants came hurrying up, evidently fully informed of the circumstances of the case, for they made directly for their imprisoned comrade and soon set him free. I do not see how this action could be instinctive. It was sympathetic help, such as man only among the higher mammalia shows. The excitement and ardour with which they carried on their unflagging exertions for the rescue of their comrade could not have been greater if they had been human beings."

Romanes concluded:

This observation seems unequivocal as proving fellow-feeling and sympathy, so far as we can trace any analogy between the emotions of the higher animals and those of insects."

Today, even anthropomorphically inclined scientists would disagree with Romanes's conclusions, which were not regarded highly in his day. His peers in the budding field of comparative psychology, which followed the Cartesian view, were developing a keen interest in conducting animal behavior experiments under new, rigid scientific standards. The anthropomorphizing of Romanes created a backlash among the psychologists, who believed the study of animal behavior needed a more solid scientific foundation. As a result, Darwin's ideas regarding the expression of emotions were ignored and Romanes's collection of anecdotes hotly criticized. By the end of the nineteenth century animal behaviorists had abandoned the approach of naturalists—the observation of natural behaviors in the wild—and focused on experiments on captive animals under controlled conditions in laboratories. In 1894 the comparative psychologist C. Lloyd Morgan introduced the principle widely used and known today as Morgan's canon. It states: In no case may we interpret an action as the outcome of the exercise of a higher psychical faculty, if it can be interpreted as the outcome of the exercise of one which stands lower on the psychological scale.

Morgan's canon has been associated traditionally with Occam's razor, from the fourteenth-century philosopher who said, "Entities should not be multiplied unnecessarily." This means keep things simple when possible, and is synonymous with the law of parsimony, a rule in science and philosophy that says "the simplest of two or more competing theories is preferable" and that "an explanation for unknown phenomena should first be attempted in terms of what is already known."

These are all solid principles for investigating scientific questions, but with the prejudiced view that animals are devoid of cognitive function and emotions, intelligence and emotion were stripped from any discussion of animal behavior. Adhering to the most parsimonious explanations, scientists must interpret behavior as physiological reactions. Leaders of the field of comparative psychology—although not Morgan, for reasons I'll mention in a moment—became strict enforcers. Anthropomorphic descriptions of animal behavior were blasphemy to science.

According to E. O. Wilson, Morgan's canon ushered in an era of reductionism—*reductio ad absurdum,* Wilson called it. Behavior that Darwin would have described as "frightened" in an animal had to be described under Morgan's canon in more clinical terms. But some scientists in the field today are saying, Enough. Based on new understandings of evolution, genetics, and physiology, they believe the interpretation is extreme. Marc Bekoff, at the University of Colorado, put together a unique, rather daring book called *The Smile of a Dolphin.* Bekoff gathered anecdotes about animals from leading animal behavior and communication scientists in which emotions were difficult to ignore. Christine Drea, an assistant professor of biological anthropology and anatomy at Duke University, who studies hyenas, provided an example of the extremism in a passage from *The Smile of a Dolphin.* Drea had returned to a facility where a group of hyenas are studied at her university and discovered that a female had savagely attacked one of the males. The male was cowering and shaking with fear. Females are higher ranking than males in hyena society and they can be quite the doms. This is how Drea says one must describe fear when it is observed in an animal: "In scientific terms, he was a low-ranking hyena who had suffered the stress of acute changes in circulating cortisol concentrations brought on by social interactions with higher-ranking animals. In laymen's terms, he was merely a frightened hyena who needed comforting."

When Morgan presented his principle in 1894, it was immediately interpreted as a criticism of Romanes and a rejection of anthropomorphism. Whether Morgan actually intended such criticism, and whether he meant for his rule to be interpreted so rigidly, is a matter of recent debate. According to Roger K. Thomas, professor emeritus of psychology at the University of Georgia, Morgan was interested in Romanes's anecdotes and enjoyed them. Morgan reportedly had said on at least one occasion that he did not see how animal behavior could be described without anthropomorphizing to some extent. Thomas's research on Morgan found that he had tried in 1904 to have his principle interpreted from a less sterile perspective. Morgan apparently viewed Romanes's anecdotes as interesting observations of sometimes remarkable animal behavior that could be explored further and tested. Morgan, in his own book, *The Animal Mind,* explained how anecdotes should be handled by sharing one of his own about the ability of his fox terrier, Tony, to lift the latch to the front gate and let himself out into the street "where there was much to tempt him—the chance of a run, other dogs to sniff at, possibly cats to be worried." Morgan had become curious about how the dog first learned to lift the latch.

Had the dog worked it out like a puzzle to be solved, or was there another explanation? Morgan discovered the answer only by observing Tony over time.

As it happened, Tony would put his head between the bars of the gate to "eagerly" watch the comings and goings out on the street. On one occasion, Tony placed his head by chance beneath the latch. When the dog raised his head, he accidentally lifted the latch and the gate swung open. After about a dozen different occasions, Tony learned by association that if he stuck his head between the right set of bars and raised it up, the gate would swing open. Learning by association, as in the case of the blue tit and the milk bottles, is common in the animal kingdom as well as in humans. Morgan said that after Tony had learned the behavior, no amount of biscuits could persuade the dog not to unlatch the gate. Morgan wrote in his conclusion to Tony's story:

Tony, the fox terrier, belonging to C. Lloyd Morgan, learns to let himself out into the street by placing his head under the bar. Morgan solved the mystery of Tony's new trick.

> What I wish to stand out is this: that in some sense of the words Tony had, as I believe, an end in view, or "objective," namely to get out into the road; but that he had not at the outset, before gazing out there . . . discovered the means by which this end in view could be attained. Now the outcome of the completed behavior, lifting the latch and getting out into the road, was such as anyone who chanced to pass by might readily observe. I sought to ascertain through what

progressive steps this outcome was reached. The point here is that observation on one occasion only, no matter how careful and exact that observation may be, does not suffice for the interpretation of this or that instance of animal behavior.

Morgan's words do not sound like those of a man who believed that animal behavior should always be interpreted as a physiological response. They sound more like the words of a cautious, thoughtful scientist who believes anecdotes ought not to be overinterpreted.

Even so, the story of Clever Hans, the intelligent horse, is the classic cautionary tale taught to students for most of the twentieth century about the perils of anthropomorphism. The results of an investigation into the story were published as a book in 1907 by a young German psychologist, Oskar Pfungst, who proved the horse to be "dumb" after all. A growing number of scientists would interpret the outcome differently today. But first, the story:

Clever Hans was a Russian trotting horse owned by a retired German mathematics professor, Herr von Osten, early in the twentieth century. Clever Hans and Professor von Osten did not converse in a Dolittle fashion, but Hans appeared to possess the ability to perform mathematical calculations, spell words, recognize colors, and distinguish between different types of music. Professor von Osten spent a lot time working with Hans at a chalkboard to give the horse its lessons, much as he had done with his own students at the university. Hans would respond to his teacher's questions by tapping his foot the appropriate number of times for the answer to a math problem, or by nodding his head, or by picking up objects in his mouth or pointing his nose.

Clever Hans's answers were correct most of the time, as long as the questioner was von Osten or someone familiar. The professor was quite impressed with Hans's apparent brilliance as well as his own ability to teach Hans complex human concepts such as math and music theory. Von Osten was not interested in exploit-

Clever Hans became a popular attraction among members of German aristocracy. Hans's innate and remarkably acute perception made it appear that he could solve mathematical problems and understand music. Horses and other large mammals are masters at reading subtle changes in body language and other nonverbal cues to communicate. For a century, scientists have generally failed to appreciate Hans's natural talents, teaching students not to overinterpret animal behavior and to guard against cuing during behavioral research.

ing Hans for money, but simply enjoyed inviting other professors and members of Berlin society to his estate to observe Hans's mental prowess.

As word spread about Clever Hans, von Osten found himself embroiled in a controversy and accused of defrauding his invited audiences. Because scientists at the time were sure that animals did not possess any ability to think or make decisions, a panel of experts was commissioned to get to the bottom of this nonsense about a counting horse with human-seeming mental abilities. The 13-member investigating panel included the comparative psychologist Carl Stumpf, the founder of the Psychological Institute at the University of Berlin, who was determined to show that no animal could possess mental reasoning or any motivation beyond that of obtaining a carrot as a conditioned response to a cue. In Stumpf's view Hans was no more than a trained circus animal, and von Osten a con.

The scientific panel set about to discover whatever cues or signals von Osten was providing to Hans to obtain the seemingly correct answers to all of these questions regarding math and music theory. The professor was greatly embarrassed, but after five days of the inquisition, the panel came up empty-handed. Stumpf was stumped. After he returned to Berlin, Stumpf assigned the frustrating case to a young student, Oskar Pfungst, who returned to von Osten's estate to conduct his own study. With a cleverly designed approach in which he varied the conditions under which Hans was tested, Pfungst was able to show that Hans was indeed reacting to cues, however unintentional they might have been on the part of von Osten and other questioners. Hans was clever, but not in the way von Osten believed. Pfungst was able to show that Hans had no concept of the questions being asked, since they could be asked in a foreign language, whispered in an almost inaudible voice, or delivered telepathically. No matter how a question was posed, Hans would somehow arrive at the correct answer. The break in the case came when Pfungst observed that Hans's accuracy began to fall dramatically after dusk. Pfungst became certain that Hans needed to see the questioner or at least members of the audience who knew the answer to the question. If the questioner stood behind an opaque screen, and Hans had no other informed person to watch, the clever horse appeared as dumb as Pfungst expected.

Pfungst was convinced that von Osten had been lying to him and the investigating panel all along, but there was one snag. Even though Pfungst knew Hans was reading cues, the psychologist could not prevent himself from delivering the correct answers to Hans during his own tests. It turned out that Hans was just doing what horses and wild animals do naturally: exercising an acute perception of body language and emotional states. Hans was listening to the tone of the questioner's voice and paying attention to the most subtle of eye movements, shifts in posture, breathing, and facial expressions to determine when to stop nodding or pawing or whatever the test called for him to do. If someone asked Hans to add 5 plus 5, he would paw the ground while carefully observing the questioner. In the split second between 10 and 11, Hans was able to perceive some subtle relaxation of the questioner's body, a twitch of the eye, a rising anticipation in the voice, or some other reaction that human perception could not detect.

What was unappreciated at the time and for most of the past century was that Hans was doing a remarkable job of communicating with a species other than his own. Hans could not do math or understand music theory, but there is no reason that a horse should have evolved such abilities. Neither would have been useful when competing with stallions for mares or sensing predators. Hans's perceptiveness, and that of all horses, has been favored and honed through natural selection to improve the chances of survival.

Pfungst and Stumpf rejoiced in Hans's defeat and in their ability to prove that animals are unable to think. Professor von Osten was so disillusioned by the experience that he reportedly sold Hans to an animal foundry. If it is any consolation to horse lovers, Hans bit Pfungst twice quite deliberately during the study.

Had the comparative psychologists appreciated the genuine intelligence that Hans possessed, and not been so rigid in their interpretation of Morgan's canon, research on animal behavior might have been more productive over the past hundred years. This is not to say great strides have not been made, but perhaps cruel experiments, such as rearing infant primates in isolation, could have been avoided. Admitting that animals possess intelligence, emotions, and even cultural behaviors surely would have influenced how the scientific community treated animals in laboratories and pharmaceutical studies.

Change has been slow and relatively recent. Studies of animal cognition have accelerated most dramatically in the past ten years, though a few pioneers have been at work since the 1970s. Cases such as Betty the crow and many others demonstrate beyond a doubt that animals are intelligent, communicate with intent, and have the ability to solve problems. Natural selection favors animals that can remember events and use that information to make decisions in the future. Donald Griffin, the professor emeritus at Harvard University who discovered bat sonar, has said that animals should be regarded as actors in life's grand play rather than passive objects. According to Griffin, animals in general have evolved "a simple but conscious and rational thinking about alternative actions" and the ability to choose the actions that will get an animal what it wants or avoid what it fears.

It was Griffin, perhaps more any other scientist, who prodded the scientific community to begin accepting that animals are thinkers, communicators, and problem solvers, and he is regarded as a founder of the field of animal cognition. His involvement certainly helped many younger scientists who held similar views to speak up. In 1992, Griffin published *Animal Minds,* which was based on a broad collection of studies by other scientists demonstrating the varying cognitive and communicative abilities of animals and was written primarily to get the attention of other scientists. Griffin made the case that animal communication provides a "window" into the animal mind. The way a species communicates reveals its cognitive capabilities and the extent to which it is able to exercise nurture over nature. Having some flexibility within one's hardwired behavior allows an animal to adapt more quickly to changes in its environment, to assess situations and invent or adopt new behaviors. All are qualities that natural selection would favor, but for at least the first half of the twentieth century animal communication was essentially the unwanted stepchild of animal behavior research. Griffin wrote that "psychologists have paid little attention to the communicative behavior of animals, for reasons that are not entirely clear. Could this lack of interest stem from the pervasive inhibitions of behaviorism for the very reason that communication does suggest thinking?" According to Griffin, communication was regarded as the expression of internal physiological states that involved no conscious thought, such as blinking of the eyes and expressions of pain. Vocal and visual signals were viewed, as Descartes wrote 400 years ago, as "involuntary movements." Griffin called this the groans of pain, or GOP, interpretation of animal communication.

Among the leading examples Griffin uses to illustrate the association between conscious thinking and communication are the semantic alarm calls of vervet monkeys and the semantic screams of rhesus macaques. Alarm calls, like those of the douroucouli monkeys of the Amazon, are common in the animal kingdom, but the calls of vervets appear to represent a rudimentary form of language. Vervet monkeys live on the savannas and in river woodlands in East Africa. They are medium-sized and form troops of 20 to 60 individuals.

Their alarm calls were first described in 1967 by Thomas Struhsaker, a research scientist in the Department of Biological Anthropology and Anatomy at Duke University. Struhsaker, who is a leader in the study of primate communication, suggested that the vervets were making different alarm calls in response to different predators and that the calls appeared to prompt specific responses. Struhsaker believed the calls were adaptive—they had developed in the vervets through natural selection. This work stimulated much interest and controversy among animal communication researchers. It was not until ten years later, in the summer of 1977, that Robert Seyfarth and Dorothy Cheney of the University of Pennsylvania, and Peter Marler, now professor emeritus at the University of California at Davis, all pioneers in animal communication, traveled to Amboseli National Park in Kenya to record the vervet alarm calls and conduct playback experiments.

"Do nonhuman primates in their natural habitat signal about objects or events in the world around them? A methodological problem confronting any observer who attempts to answer this

A young black howler monkey will soon become the dominant member of his fiercely territorial family. Howlers, which populate the few remaining Central American rain forests, declare their presence each morning at dawn with loud calls that can be heard more than a mile away. The frequency of the territorial calls varies with the size of the monkey. A chorus of calls of varying frequencies lets neighbors know how many family members, and sometimes unrelated allies, are ready to defend the homeland.

question is that an animal cannot be interviewed. Instead the observer must try to arrive at the content of each signal by studying the responses it evokes in other individuals," Seyfarth and Cheney wrote in a review of their research.

Their careful studies, conducted in 1977 and again in 1978, demonstrated that vervets use alarm calls with specific meanings to warn each other about different kinds of predators. The vervets' primary predators are leopards, martial eagles, and pythons. One call means "Leopard!" Another warns "Eagle!" A third says "Snake!" When the leopard call is given, the troop runs for the treetops. When the eagle call is given, they head for the ground. The snake call prompts them to stand up and look around. Applying the law of parsimony, the scientists needed to determine whether the calls actually served as words that symbolically represented each predator, or whether they simply reflected varying degrees of emotional arousal as a response to sighting a predator.

If the calls refer to the predators, similar to the signifying of words, then the vervets should react consistently to the calls when they hear them, even if no predator is in sight. By playing the recorded calls under controlled conditions over speakers in the wild, the scientists showed that the monkeys reacted the same way they did to a live call and that they did so in the absence of a predator. The calls of the vervets are genetically fixed through natural selection, but their proper use takes some schooling in the young. Juveniles make eagle alarm calls in reaction to harmless birds and even falling leaves. Likewise, they make leopard and snake alarm calls in response to anything moving on the ground. The adults are aware that the young vervets are learning and usually just glance at the young caller and go about their business. What is noteworthy is that young vervets are distinguishing between classes of objects, even if the calls are not specific. Only humans were thought to have the ability to categorize objects in the environment, but vervets use the eagle call for objects in the air and the other calls for objects on the ground. Furthermore, the young vervets learn to make the calls in the appropriate contexts by observing the adults. Seyfarth and Cheney find parallels with infant development of speech.

"Human infants learn the association between specific words and specific objects or events in a complex social environment, where cues from other individuals play an important developmental role," Seyfarth and Cheney said. Apparently, so do vervets. When a young vervet utters the correct alarm call in the presence of the actual predator, the adults also make the call, reinforcing the proper association.

Vervets also make calls known as chutters and *wrrs,* which are used in different contexts. The chutter, given by a member of the troop to warn the others that another vervet troop is approaching, has the effect of calling everyone home. The *wrr* also appears to alert troop members that another group has been sighted, but it carries less urgency. Experiments done with these calls demonstrated that troop members recognize each other's voices and recognize each other as individuals. In these experiments, the chutter of one member was recorded and played back to the troop repeatedly. Initially, the troop responded as they normally would, by gathering together. But after about the eighth time a call was played, the troop caught on that no other groups were around and ignored the caller's voice. When the *wrr* of the same individual was played, the group also ignored it, indicating that the group knew which monkey was responsible. Playing the chutter of a different individual, however, caused the group to respond normally.

The females also appear to recognize the voices of infants and understand which infants belong to which mothers. When playbacks of a particular infant's distress calls were conducted, the mother looked toward the speaker, while the other females looked toward the mother. Seyfarth and Cheney found that vervets recognize each other as individuals, recognize family members and members of other families, and understand their place in the dominance hierarchy of monkey society. "The ability of monkeys to classify individuals in this manner is almost certainly the result of natural selection acting on animals with a complex social framework, where detailed knowledge of relations among all members is essential," they said. Once the ability to classify group members becomes fixed in the species, the monkeys can use it to classify objects in their environment such as eagles, leopards, and snakes. This ability can be applied to more subtle social settings, too. Vervets use a graded series of grunts that are context specific and individually distinctive, and that in some cases refer to external events. These abilities appear in other species of monkeys, including macaques. They may be fundamental to the origins of society and a necessary component for the origin of human language.

Harold and Sally Gouzoules are primate researchers based at Emory University in Atlanta. Since the 1980s, they have been studying social relationships among rhesus macaques transplanted decades ago from captive environments at pharmaceutical companies to Cayo Santiago, off the coast of Puerto Rico. They conducted studies, with Peter Marler, that identified semantic screams used by macaques during social conflicts. They discovered that macaques have five distinct types of screams: noisy, arched, tonal, pulsed, and undulating. Females establish dominance hierarchies, with offspring acquiring status just below that of their mother. Among the offspring, the youngest ranks highest, followed by the next youngest. Conflicts often arise between young macaques. Dominance hierarchies in nonhuman primates resemble human office politics. Bosses come and go, and lower-ranking individuals can claw their way to the top. Occasionally, in macaque society, a lower-ranking female will sneak up and deliberately slap a high-ranking female's offspring when the mother is out of sight. If the mother does not respond for some reason, the lower-ranking female climbs a rung up the social ladder.

The semantic screams are uttered by an individual that is under attack. The scientists discovered that the type of scream identifies the status of the monkey delivering the beating or creating the conflict. Noisy screams are normally used when the attacker is higher ranking and making physical contact. If the high-ranking opponent is only threatening, undulating screams are given. The macaques make arched screams during nonphysical contact with lower-ranking attackers. When the attacker is a sibling or other relative, tonal and pulsed screams are given. Mothers respond as one might expect.

"If the attacker is higher ranking and the aggression is strong, the mother comes running and will recruit other allies that are usually related. She will come in and slap the attacker and run off, or approach and scream bloody murder to distract the attacker," Harold Gouzoules says. "If the attacker is lower ranking, the intervention could be by the mother or a relative and it will be bold and confident. It is good to have this information about the attacker before appearing on the scene. The fact is, they don't always intervene. These calls are adaptive from the ally's perspective as well. They need to think, Hmmm, do I want to come in and help my cousin today or not?"

Deception is another of the Gouzouleses' interests. Primates have hierarchies among both males and females. Dominant males and females often get first servings whenever a foraging troop sights a food cache. Males at the bottom of the social ladder have to stand by and wait for the leftovers. Anyone raised in a large family will find that scenario familiar. Macaques generally forage out of sight of each other so when one member of the troop sights a bunch of ripe fruit on the ground, it will give a food call. Natural selection has seen to it that the food call is hardwired so that everyone gets to share. Primates, including many humans, simply cannot suppress food calls no matter how much they might want to keep quiet and hoard dinner for themselves. Reporters know that free cake is available from the excited squeal that someone always emits. Reporters recognize the sound, and even if it is near deadline, they leave their desks and make for the cake.

Harold Gouzoules found that subordinate macaques may arrive quickly on the scene of a food find, but as soon as the dominants show up, they are booted to the periphery. Then comes the deception. "On a couple of occasions I have seen a low-ranking animal suddenly produce an alarm vocalization. All those high-ranking animals run for the trees, and what does the alarm caller do? It sits there stuffing its face. Now, do they do it on purpose? If so, why don't they do it more often?"

Gouzoules followed up with more recent studies, showing that the monkeys quickly catch on to the "boy who cries wolf." He played the alarm calls of higher-ranking monkeys to a group, observed their reactions, and then played the alarm calls of lower-ranking members to the group. The group always reacted to the calls of high-ranking monkeys, even when there was no evidence of danger, but they quickly stopped reacting to the calls of lower-ranking monkeys once they saw there was no evidence of danger to back up the calls. "That requires a sophisticated level of cognition," Gouzoules says. "What they are saying in response to these calls is that 'You are so-and-so at the top and have nothing to gain by lying, while you are so-and-so and have much to gain.'"

Life in the wild is an arms race in many respects. Deception is rampant in nature, having evolved through natural selection as a means of survival. Much of it is not done consciously. For instance, some flies look like bees to deter animals from eating them, hognose snakes play dead, and animals' colors and coats blend into their habitat. Among social primates, however, deception does appear to be willful to some extent. As a countermeasure, to ensure that primate society is not overrun with liars and cheats, skepticism evolved. Studies have confirmed that members of a macaque troop trust the calls of their leaders but are wary of those of lower-ranking members.

These types of studies reveal that many qualities and behaviors traditionally regarded as human have deep evolutionary roots in the animal kingdom. So perhaps we should not be too critical of those twentieth-century skeptics about animal cognition and emotions. They were just acting on their primate instincts.

Mark Hauser, who studies primate cognition at Harvard University, believes that all animals are equipped with a "set of mental tools for solving social and ecological problems." Hauser's work suggests that some of the tools are universal and provide fish, insects, reptiles, birds, and mammals with a fundamental ability to recognize objects, know the difference between more and less, and navigate. Differences in abilities emerge as species encounter different social and ecological problems. Long-lived social animals, such as primates, elephants, and marine mammals, have evolved a more deluxe tool kit than other species. Now it is time to explore exactly how animals have come to acquire these tool kits and abilities.

What do scientists mean when they talk about evolution? Griffin referred to it as life's grand play. The show is about to begin on Midway Island, in the Pacific Ocean.

Three

THE SHOW MUST GO ON

AT 5 A.M., the narrow paved streets that lead from the old officers' quarters to Cross Point observation deck on Midway Atoll are still wet from a late February thunderstorm that blew over shortly before midnight. As I cycle the streets at this predawn hour, a thin line of spray from the rear tire forms on the seat of my pants. Minor congestions of busy field mice, brazen little beasts, block my progress, scurrying out of the way only at the last minute and forcing me to brake and zigzag to avoid squishing them. The sun does not peek over the horizon until 6:30, so it is relatively dark except for the stars overhead and the halos of mercury vapor streetlights that dot the tidy intersections of this former U.S. Navy base. This is one of the rare times when juvenile albatrosses are not strutting, bobbing, weaving, clapping, flapping, and rapping in vocalizations of teenage courtship. A lot them carry on until 3 a.m. before giving themselves and the few people here some peace.

Cross Point faces east and is a good place to get a seat for the grand show that begins at sunrise. The modest observation deck and white wooden cross overlook a short abandoned pier at the point, which becomes part of the stage. The albatross season on

Midway has not changed dramatically over the past 4 million years. Nature, the director, does not tinker with the basic plot or tunes of a show once a species has proved itself a success in one of its performances. Based on its long run at Midway, the show starring the island's 1 million albatrosses is a smash hit.

Midway Atoll is located 300 miles north of the Tropic of Cancer and 1,250 miles west-northwest of Hawaii. The atoll consists of three small islands, a large central lagoon, and a barrier reef that supports abundant numbers of tropical fish, many that avid scuba divers in the waters around Hawaii may see only once in a lifetime. Residents include 17 species of seabirds, 200 spinner dolphins, and a small number of endangered monk seals trying to make a comeback. The leading attractions of the atoll are the albatrosses, which lie sleeping on the manicured lawns of Sand Island, where 150 people also reside, on the cracked and abandoned runways directly across the lagoon on Eastern Island, and on the only strip of natural habitat, called Spit Island.

Settled in 1905 as a way station for laying telegraph cables across the Pacific, Midway is better known for the World War II battle between the United States and Japan. On arrival, people immediately began remodeling the island, importing ironwood trees for windbreaks and shade, planting lawns, erecting buildings, dredging channels in the lagoons. By the mid-1930s, Pan American Airlines was using the island as a refueling station for the Trans-Pacific Flying Clipper Seaplane and operating a hotel for its passengers. In the 1940s, Midway was converted to a strategic naval base and military airfield, and from that point until 1996, when the navy deserted the island and cleaned up its considerable mess, the atoll was an inhospitable place for anything nonhuman. People had also introduced new species, including rats, which spread from the holds of ships and nearly wiped out the population of Bonin petrels that lay their eggs in elaborate burrows. But nature kept performing, through all the stages of human presence.

In 1905, two canaries were brought over by the wife of a tele-graph executive. Sand Island now has at least three different breed-ing populations of canaries, whose evolution has resembled that of the finches of the Galápagos Islands as recorded by Charles Dar-win, and by Rosemary and Pete Grant in the second half of the twentieth century. The canaries have reverted to the behavior and songs of wild canaries and, like the finches, have established differ-ent breeding populations and song dialects depending on which stand of ironwood trees a group claims as territory. The Galápagos finches have also been shown to have evolved different beak struc-tures from island to island in response to climatic changes and food availability in the relatively short period of time (20 years) that the Grants have been studying them. They're a fascinating story of evo-lution in action. But the Midway canaries have not been studied in any detail. The atoll has only been open to the public since 1996, and the U.S. Fish and Wildlife Service, which is guiding Midway's return to nature, has had more pressing concerns than studying the canaries—such as protecting the monk seals, which raise their pups in the lagoons, and ridding the three islands of rats.

The Laysan albatrosses are the dominant species. They have peach-colored beaks, dark eyes, and are predominantly white, with black wings and tail tips. They number about 700,000, compared with 300,000 black-footed albatrosses. Both species arrive every year in late October through November. The single-minded pur-pose of albatross adults at Midway is nesting, mating, and nurturing a chick until it is almost ready to fly. Juveniles, meanwhile, rehearse and rehearse the choreographed courting ritual that will one day land them a mate and usher them into the role of parenting. In May, the juveniles and adults begin departing for their summer feeding grounds in the Arctic. Abandoned by their parents, the nestlings hang around a little longer and only leave if they are strong enough to fly. By the end of July, nestlings are either soaring on the breezes following some preordained flight plan or they are dead. To survive in this show and return for future seasons, chicks have to run a gauntlet of risks—dehydration, starvation, drowning, and predators such as the tiger sharks that patrol the beaches waiting for the weaker chicks to drop into the water.

The first, barely detectable photons of the solar alarm clock reach Cross Point at 6:15 a.m. The seabirds begin to stir and vocalize to each other with gentle "eh-eh" contact sounds. They rise, face the warm wind, and stretch out their wings. A handful of eager juveniles go straight into their dance routines, drowsily bobbing their heads and strutting in hopes of finding a willing partner. With the wind blowing from the north, meeting the end of the old pier head on, the albatrosses have a perfect runway. Nicknamed the gooney bird, an albatross is a remarkably graceful, highly efficient long-distance flier. It sometimes has a little trouble lifting off and landing—a design trade-off made by the species long ago. More than a few can be seen walking around with dirty brown stains on their breasts made from crash landings.

The horizon has become a thin red line, and the sky above it is shifting from black to purple. A group of albatrosses approach the pier, lining up for takeoff like the planes at Reagan National Airport in Washington, D.C. The first flier races down the runway, webbed feet slapping loudly on the cracked pavement as it waddle-runs and unfurls its wings. The bird uses up about three quarters of the pier's length to build enough speed to fill its nearly seven-foot-long wingspan and gain loft. One by one, albatrosses follow each other down the runway and take off into the wind. The eager juveniles, having struck out at finding dance partners, join the others lifting into the air. At dawn, just about every bird at Cross Point is gliding along the upwelling drafts created by the stands of ironwood trees that line the small inner harbor of Sand Island.

After the birds have landed and the sun has risen above the horizon, the courting begins. The air comes alive with the rat-tat-tat clapping of beaks, high-pitched whinnies, soft eh-eh's, honks, and celebratory "sky moos" that are the repertoire of the courting juveniles. During the final sky moo, the albatrosses crane their necks, pointing skyward, and moo together. The adults, meanwhile, will take turns feeding out at sea and returning to the nests, which are scooped-out depressions in the sandy soil with a little dried grass on the bottom. Spaced only three to four feet apart, they cover the entire island. The parents regurgitate a nutritious oil made from partially digested squid or fish into the chick's open mouth. This is daily life on Midway.

Albatrosses are highly vocal with each other and do not hesitate to express their displeasure at any intruder who approaches a nest too closely. The adults clap their beaks aggressively and hiss, warning me that I am trespassing and should step away from their nest and its offspring. Their next communication will be to utter harsh honks and charge. If forced to, they will bite with their four-inch-long beaks, which can inflict impressive gashes, one of which I've seen on the hand of a Midway ranger. Albatrosses squabble among themselves quite a bit, but during the nesting season nothing gets much more serious than a cranky honk and an occasional nip on another's beak.

I did observe one fight break out, however, between two dance partners. Males and females are nearly the same size and perform almost the same moves. This makes it difficult for scientists to determine the sex of a bird without a closer inspection that might get their hand severely bitten. Apparently, albatrosses may occasionally get a little confused about the sex of the bird they're dancing with, too.

Sitting in a field on Sand Island near the old Victorian telegraph houses, I watched one albatross change partners eight times in less than half an hour. The albatross, which I had named Bob, was advertising its availability for a partner by bobbing its head and strutting. Another albatross walked up to Bob, bobbing and strutting, and they began to engage in the ritual moves. They were clapping their beaks and moving their heads like Mick Jagger when suddenly they both stopped and attacked each other rather viciously. It seems two males had discovered that they were dancing with each other. Each stomped off and shrugged its wings and appeared embarrassed, though this sounds like something George Romanes would have said. Clearly, they were upset.

Male and female juveniles also get upset when a dance ends before they've gone through all the steps. They clap their beaks at each other when one makes a wrong move as if to say, "Honestly! Don't you know how to dance?" They are highly motivated to learn their steps, and all appear to be serious students. Some are naturally better than others and will graduate from training and secure a mate within a few years. Most will require up to five or six years of practice, but a few wallflowers—males—may not succeed for ten years or more.

Depending how one analyzes the choreography, the courtship dance has anywhere from 8 to nearly 20 steps. During a successful performance, a male and female first approach each other, bob their heads, and touch beaks. The couple slowly circle each other, and each vocalizes with an appeasing, gentle-sounding eh-eh. They perform a variation of a step called the flight-intention movement, in which they partially raise their wings but do not unfold them. They alternate clapping their beaks in rapid succession, which creates the rat-tat-tat sound. Both shake their heads wildly and whinny. Toward the end, they alternate craning their necks toward the sky and release the joyful-sounding moo. If all goes well and the partners reach the end without making any serious mistakes, each pulls up blades of grass with its beak, which symbolizes nest building, and they sit down together as if they were already mates. This lasts for only a few hours before they move on and choose new dance partners, a bit like a middle school romance. The courting rituals carry on from dawn to long after dusk, and the beak clapping, honks, whinnies, and sky moos are heard everywhere one goes on the small island.

Ritualized dances are a form of visual communication. Some of the albatross's ritual movements are similar to those found in unrelated species, such as the head-shaking ceremony of great crested grebes, and the symbolic nest-building ritual at the end of the dance, when grebes present waterweeds to each other. Flight-intention movements are quite common in all types of birds and may have different meanings depending on the context. The similarities reveal a good degree of conservation in the choreography of routines across species. That may be the result of having distant common ancestors, or it could be that different species have arrived at similar solutions to similar problems.

Communication begins quite early in the albatross's life, as it does in most species, between a parent and its offspring. When an albatross parent, male or female, approaches its nest, it lowers its head and utters a soft greeting eh-eh at the egg. The call probably serves to orient the offspring to its parents' voices after it hatches. Penguins, which colonize in large rookeries, do the same thing. Both mammals and birds commonly use vocal signals for parent-offspring recognition. In some cases, a young bird inside an egg will vocalize to its parent when its body temperature falls, which encourages Mom or Dad to sit back down and warm it up. Becoming oriented to parents' voices is critical for species that nest in such large colonies. If a chick wanders off from the nest and becomes lost amid the chaotic comings and goings of a million albatrosses, it will not be fed and might try begging. When the parents are away, the fuzzy, awkward nestling is left alone for hours, sometimes days. Neighboring parents have been known to mug a nestling when its parents are away at sea feeding. Mugging a defenseless chick seems harsh, but it appears to be an albatross's way of discouraging a neighbor's chick from crashing someone else's dinner in the event that it is orphaned. There is no place for orphans in albatross society. Commercial fishermen's long lines, which float near the surface for miles behind a vessel, have become a frequent killer of foraging albatrosses. Parents spend most of their time and energy, and risk danger, gathering resources for one chick only—their own.

Adults find their way back to their own nest among the 400,000 that are spread over Midway Atoll's 2,000 acres by using mental maps, navigational landmarks, and a sense of smell—common tools among many species, including bees. Adult albatross pairs return to the same nesting site, give or take a couple of feet, every year for the rest of their lives. The males arrive first, usually the male black-footed albatrosses, to stake out their real estate and defend it from young interlopers. Older breeding males take this job very seriously, and the squabbling, which is sung in low, cranky tones, becomes quite aggressive in October and November.

Even after Midway was bombed by the Japanese in World War II, and after the U.S. Navy tried in 1957 to rid it of the birds by paving the islands with concrete and bulldozing the nests, the show kept right on schedule. The albatrosses continued to arrive at the same time of year, mating, building nests, raising chicks, and courting with the traditional dances despite the significant resources the U.S. military employed against them. Four million years of genetic programming in albatross behavior was a formidable force. The navy finally gave up and today the albatrosses glide on the breezes at sunrise.

The force of nature, of life, is humbling, and often mysterious. It operates with an amazing consistency and predictability, using the same basic script in every environment no matter what species is involved in the performance. The plot is about surviving and finding a way to mate in order to pass along one's genes. Even in such highly social species as primates, the story boils down to the individual, as it does in Ayn Rand's *Fountainhead* and *Atlas Shrugged*.

Genetics determines the basic behavior of both animals and humans to a large extent, but nature has made room for learning as species faced new challenges. Built into most animals' genetic script is a degree of variability that allows for individuals to be flexible enough to adapt to challenges during times of both great and subtle change. In a sense, the actors are allowed to rewrite or add new lines to their genetic scripts when needed to fit the circumstances.

Research conducted by Lee Dugatkin, a professor of evolutionary ecology at the University of Louisville, and a number of other scientists has shown that learning plays a large role in animal behavior in species as simple as guppies. Dugatkin's book *The Imitation Factor* explains the evolutionary underpinnings of learning, which is important in many areas of animal communication, including birdsong and marine mammal vocalizations. Animals, just like humans, are able to learn new behaviors during their lifetime and pass them along to succeeding generations by means of cultural transmission. *Culture* has been used traditionally to define human behavior, but cultural behavior appeared in animals long before humans arrived on the scene. Cultural transmission of behavior in animals is a relatively new field of study, and its findings are only now being incorporated into the understanding of Darwin's original evolutionary theory.

"Biologists who study evolution primarily thought of behavior as being under some direct sort of genetic control. That control could be very complicated, but it would be viewed in terms of how selfish genes code for behavior. Now we know that imitation and other related kinds of phenomena also have a significant impact on animal behavior," Dugatkin says.

One of the chief ways cultural information is transmitted in humans is through imitation. Any parent knows too well that children imitate adult behavior. In the third century B.C., Aristotle wrote: "Imitation is natural to man from childhood, one of his advantages over the lower animals being this, that he is the most imitative creature in the world, and learns at first by imitation." But according to Dugatkin's research, "Aristotle was wrong, at least for the most part, about the ability of non-human animals to imitate. It was a big oversight."

Results of Dugatkin's experiments with the lowly guppy indicate that imitation is endemic to the animal kingdom. It's no secret that females and males of the human species find certain traits in the opposite sex more attractive than others. Popularity and good looks often top the list. Studies at the University of Louisville found this to be true in students and further revealed that if one person wanted to date X, so did others—a phenomenon called "date copying." Dugatkin designed an experiment to see whether guppy love follows a similar path. As a rule, female guppies prefer to mate with bright orange males. The preference is determined by guppies' genes. In Dugatkin's experiment, he placed female guppies in an area of their fish tank where they could observe bright orange males and dull males at the same time and also see a female making a mate choice. Through a trick with opaque and see-through dividers, the females observed what appeared to be another female selecting a dull-colored male for mating. Surprisingly, the female observers also then chose a dull-colored male, overruling their normal instinct to mate with brightly colored males. Mate choice copying has been found in other fish species, Japanese quail, and small marine animals called isopods.

Frans de Waal, a primatologist at Emory University, became convinced through his work with chimpanzees that strict genetic programming cannot account for many types of behavior seen in animals. Instead, genes have evolved to provide an open playbook of sorts in which some animals can write new behaviors. These are learned through observation and imitation. Interestingly, neither Dugatkin nor de Waal was aware of the other's work until recently. De Waal has described his research and that of other scientists in *The Ape and the Sushi Master,* drawing his title from the apprenticeship of a sushi chef. The sushi apprentice in Japan performs menial tasks in the master's kitchen for several years but never touches fish. At the end of the observation period, the apprentice is expected to prepare sushi as perfectly as his master.

"There is growing evidence for animal culture—most of it hidden in field notes and technical reports—that deserves to be more widely known," de Waal says. "Culture simply means that knowledge and habits are acquired from others, often, but not always from the older generation, which explains why two groups of the same species may behave differently."

The ritualized courtship dances of albatrosses are clearly programmed by their genes. The dances are under the control or influence of evolution. Each step of the courting dances of the two species of albatross on Midway is predetermined and invariable, but the dances of the two species vary in a few steps. A black-footed albatross performs a full bob of its head at about step four instead of clapping beaks as the Laysan does, and it spreads its wings wider at step seven. The occasional hybrid chicks of male black-foot and female Laysan pairings have moves that are intermediate between the two species—instead of the full head bob of a black-foot, the hybrid does a half-bob and moves its wings half as much, demonstrating unequivocally that the moves are genetically determined. The juveniles still require several years of practice at the steps, however, before they are proficient enough to select a mate. The first year that a juvenile returns to the atoll is spent observing older courting juveniles—like a freshman at a high school dance. They will try out a few clumsy steps of their own, but few will become experts for several seasons.

Both genetic programming and learning are products of evolution; both play a role in a species' communication. As Darwin developed his theory of evolution, published in *The Origin of Species* on November 24, 1859, he noted that "favorable variations would tend to be preserved, and unfavorable ones to be destroyed. The result of this would be the formation of new species." Darwin particularly sought to understand the force that drives the process by which species develop over time, thrive or disappear, and change in physical appearance from earlier forms. He derived insight from observing the common and ancient human practice of breeding plants and animals. No one questions the power to breed roses, corn, horses, cattle, pigs, and dogs and cats by selecting preferred traits and breeding only those plants and animals that possess the traits. Darwin called this common practice domestic selection. He and others suspected that the remarkable diversity of species and habits in the wild might be due to the phenomenon of selection, and he wrote in his autobiography: "I had overlooked one problem of great importance . . . This problem is the tendency in organic beings descended from the same stock to diverge in character as they become modified. That they have diverged greatly is obvious from the manner in which all species can be classed under genera, genera under families, families under suborders and so forth . . . The solution, as I believe, is that the modified offspring of all dominant and increasing forms tend to become adapted to many and highly diversified places in the economy of nature."

Variation allows species to interact flexibly and more effectively with members of the opposite sex and with members of other species of plants and animals. All manner of events and circumstances can apply pressure that forces an individual to adapt to a change. These can include, but are certainly not limited to, chaotic cycles of weather, diseases, earthquakes, or any random event that disrupts the status quo. Evolutionary pressures could arise when a stream dries up and causes a plant to disappear that members of a bird species feed on. Individuals that can find a new food source or relocate and establish new territory will survive. Animals that cannot adapt will perish. Something not appreciated until recently is that logging in old-growth rain forests can change the bioacoustic environment in which animals have adapted their vocal signals. Removing trees and undergrowth alters the way sound waves travel in a microhabitat and disrupts the orchestra. Even when logging companies behave responsibly to remove only the largest trees or a limited of number of trees, the bioacoustic environment is permanently altered and some species simply cannot adapt in time and disappear.

"Not a few migrating eastern American warblers, able to learn only one song and call in their lives, find themselves unable to adjust to the changes in ambient sound when they fly to their disappearing Latin American winter nesting grounds," Bernard Krause, a former professional musician who became a scientist, has reported. "Where these environments have been deforested, and when birds try to move to nearby and ostensibly similar or secondary-growth habitats, they discover that they are unable to be heard. Our studies are beginning to show a strong likelihood that survival might be impaired because territorial and/or gender related communications are masked."

Human-made sounds also affect animals' abilities to communicate with each other. Krause describes the sounds of the rain forest as a carefully choreographed orchestra. If a philharmonic orchestra was moved from Carnegie Hall to an open field next to a busy highway, imagine the effect on the musicians' ability to hear each other and perform. The albatrosses on Midway were able to survive Japanese bombing and the coming and going of U.S. Navy warplanes only because their communication occurs at very close ranges for a few months of the year, and they spend most of their lives at sea. Today, the major threat to albatross survival is competition for food with commercial fishing vessels.

Adaptation frequently occurs through daily struggles for existence and random opportunities that arise during the interplay between individuals and species. Plants and animals rely on each other for survival and continually affect each other's evolution. Biologists Dagmar and Otto van Helversen, of the University of Erlangen in Germany, recently discovered an elegant interplay between a vine related to the pea, the *Mucuna,* and bats that feed on nectar in the Central American rain forest. Bats navigate and search for food using high-frequency sound waves emitted from specialized vocal structures. When the sound waves return to the bat's ears, its brain interprets the sound "pictures" so it can steer away from obstacles and target food items. The *Mucuna* vine has come to rely on bats to disperse pollen the same way other flowers rely on bees. To help the bats locate the vine and its blossoms, the Helversens found it has evolved specialized flower petals shaped like tiny satellite dishes. The concave petals bounce sound waves directly back at the bat, thus making the vine easy to find in a dense jungle with lots of background filtering through the bat's ears and brain. The vine only opens these specialized radar reflectors at nightfall, when its blossoms are ready to be pollinated and bats are looking for food. Through genetic variation plants and animals continually adapt to each other over long periods of time.

The theory of evolution explains the laws of nature. Just as people can select traits to be bred in future generations of domesticated plants and animals, wild species pass along to future offspring traits that help them survive. The concave blossom of the *Mucuna* provided it with just such an advantage, and it is safe to assume that at some time in the distant past, variations of *Mucuna* that had flatter flowers had less success at getting a bat's attention. They either died out or became specialists for another species.

Moths are also a favorite food of some insect-eating bats, which use their high-frequency sound waves to target these prey. In a kind of natural arms race, moths have countered the bat radar to a degree by evolving ultrasonic hearing, which they rely on for defense and use to communicate among themselves. Moths that originally possessed the ability to sense very high frequency sound waves were better at avoiding becoming bat munchies than moths that could not hear the sounds. Survivors continued to breed more successfully than moths that were less able to detect bat radar, and because their offspring possessed the new survival advantage, they in turn had higher reproductive success than other moths in their population. Therefore most moths today are proficient detectors of the bats' echolocating sound waves.

One common misconception about natural selection is that it

pushes a species toward some ideal of perfection or greater complexity. Natural selection has no purpose. It is a force that pushes actors on the evolutionary stage to adapt and survive or die. The survivors pass their genes to another generation of actors. Darwin used the example of a wolf to describe, in general, how natural selection operates. In *The Origin of Species,* he proposes a population of wolves that have the ability to take their prey either by cleverness, strength, or fleetness. Environmental conditions or some other influence causes all prey but deer to vanish during a season when the wolves are hardest pressed to find food. Only the wolves that are slender and fast enough to chase and catch a deer will survive. Speed is favored over strength and cleverness. Muscular strength will not be favored in these circumstances, because the deer can outrun the wolves. Over time, the more slender, speedy wolves survive and give rise to wolves that are, overall, more slender and fleet of foot than in the past. The conditions could have been different and favored muscular wolves, but nature does not care whether wolves are strong or fast.

All animals are subject to environmental influences. The primary staple of the Laysan albatross is squid. Black-footed albatrosses prefer fish. El Niño, which affects water temperatures at the surface, can alter the availability of one or the other food types. Currently the Laysan population is larger, but in the past black-footed albatrosses swelled to greater numbers when the squid population dropped. The Laysan population may be beginning a downward turn, however. The evidence obtained so far suggests the decline is the result of diminishing numbers of squid, which may be due to commercial fishing.

The balance of power between species is always in some state of flux because of natural factors and more recently our own species' success as predators. Humans are a drain on nature's economy. Destruction of natural habitats, pollution, global appetite for fish, and increasing consumption of primates and large mammals in the rain forests are killing off many species.

Perhaps the most remarkable aspect of Darwin's achievements is that he did not have the benefit of modern genetics to help explain why natural selection works. A great admirer of plants and an avid observer of domestic animal breeding, Darwin understood the principles of domestic selection quite well, but he did make one key mistake. At the time, the views of a scientist named Jean-Baptiste Lamarck about how inheritance occurs were fairly popular, and Darwin accepted them. Lamarck believed incorrectly that the skills one acquires in a lifetime could be inherited. Many intelligent people today make the same mistake as Darwin—not bad company to be in.

During a visit to a research facility known as the Think Tank at the National Zoo in Washington, D.C., I met the then director, Rob Shumaker, a young scientist who worked at the time with orangutans to study how they think and communicate. Shumaker was teaching orangutans an artificial language that allowed him to study similarities and differences between great apes' and humans' ability to use language. Shumaker conducted his experiments in full view of anyone who happened to wander into his facility at 11 a.m. and 2 p.m. An older man in the audience asked, "Do the offspring of orangutans that learn to use the artificial language inherit that ability?" Darwin would have answered yes, but the answer is no. If this man's father had learned to speak Italian, German, and Japanese, his son would have not been born with the ability to speak those languages. The geneticist Gregor Mendel showed that humans and animals inherit only the qualities and weaknesses already present in their parents' genetic makeup. Mendel made his discoveries during Darwin's lifetime, but there is no indication that Darwin ever read Mendel's work. If he had, he would certainly have refined his theory.

Nevertheless, Darwin defined natural selection in *The Origin of Species* thus: "Nature, if I may be allowed to personify the natural preservation of survival of the fittest, cares nothing for appearances, except in so far as they are useful to any being. She can act on every internal organ, on every shade of constitutional difference, on the whole machinery of life. Man selects only for his own good: Nature only for that of the being which she tends."

In addition to the characteristics that are inherited from one's parents, humans and animals have individual variations that are not present in either parent. This individual variation results from the mixing of the genes during conception, a process called recombination, and from spontaneous mutations, which arise normally during cell division. The variability may or may not be important in terms of survival, but the reshuffling of genes that takes place in an offspring increases the odds that something of value may be introduced that natural selection can exploit if necessary. Some animals within a species may possess genes that make them cleverer at avoiding predators. Others may be able to resist certain parasites better than others.

The term *survival of the fittest,* which was coined by Alfred Russel Wallace, a contemporary of Darwin's who was developing his own similar theories of evolution, is used synonymously with *natural selection.* Fitness refers to the ability to reproduce, not to an animal's muscularity. Any variation in traits that offers an individual a slight advantage in passing along his genes will tend to be favored, preserved, and spread through the gene pool. Traits that are disadvantageous will tend not to be preserved.

Genes are made of deoxyribonucleic acid, or DNA, which consists of two long mirror-image strands made up of four components called nucleotides, or bases. The strands are wound together like a spiral staircase, in a molecule known as a double helix. The nucleotides pair with their opposites across the ladder to form "base pairs." The order in which these pairs occur determines codes for a gene's function.

All of life carries in its genetic script the same base pairs of coding, regardless of the species. The only difference between a lion and a mouse or an albatross and a bee is the order of their codes and the amount of coding used in the genetic script. The entire genetic code of an organism is called its genome. Bacteria have the smallest genomes. Scientists had assumed that human beings possess the largest genome, but surprisingly, although we are at the top of the food chain, our genome is not much larger than that of a mouse.

The Human Genome Project, the most ambitious endeavor in biology in the twentieth century, was completed in 2001. On a drizzly cold February 12, scientists from competing public and private efforts announced in Washington, D.C., that they had at last determined the genetic sequence, or code, of the genes that define our human species. Both teams arrived at the conclusion that the blueprint of the human body consists of 30,000 to 40,000 genes, although when they started they had assumed the number would be closer to 100,000. The common roundworm, *C. elegans,* has 19,000 genes in its blueprint, only half to two thirds of what is needed to make a human. The lowly fruit fly has 13,000 genes. Astonishingly, the human genome and that of the mouse differ by a mere 300 genes.

Having in hand the genomes of different species allows scientists for the first time to begin comparing the blueprints to see what genes different species share and how they differ. The mouse genetic code consists of 2.5 billion base pairs, compared with 2.9 billion in the human genetic code. Initial comparisons reveal that more than 80 percent of the mouse genetic code matches the human code. The major difference is the order in which that matching code is arranged, says Eric Lander, director of the Whitehead Institute/MIT Center for Genome Research in Cambridge, Massachusetts. Lander's group led an effort to find out what it is about the genomes exactly that makes humans different from mice.

"If we consider the genomes to be the book of life of the two species, they can each be divided into 350 chapters, or blocks of genes. Each species has the same chapters, but they are organized in a different order," Lander says.

The level of similarity of the 350 chapters is fairly general, but 5 percent of the coding of the two genomes is identical, which means it has been preserved throughout tens of millions of years of evolution in both species. For genes to have been preserved for such a long time, they must have very important roles in basic mammalian development. These "conserved" genes are in a very real sense a tool kit for building a mammal. Scientists theorize that mice and humans share a mammalian ancestor that possessed the same genes and may have scurried beneath the feet of dinosaurs as long as 75 million years ago.

Scientists at the Max Planck Institute in Germany have been sequencing portions of the genome of the common chimpanzee. They predict that the human genome will vary from that of our primate cousin by less than 1 percent. Efforts to sequence the entire chimpanzee genome at a variety of labs in the United States and Britain are nearly complete. No plans are underway at the moment to sequence marine mammals, but comparisons with dolphins and killer whales would prove interesting. Someday, the comparative study of genomes will bear fruit for the study of animal communication and the evolution of signals.

Scientists are actively searching for human genes that are linked to behavior. Once these genes are discovered, scientists can search for matches in the genomes of other species. Genes that perfectly match, as do the 5 percent between humans and mice, have been conserved by nature through the course of evolution. One classic example of conserved genes is the family of "homeobox" genes, which contain the coding for making the wings of a fruit fly. The same genes in a mammal contain coding for limb buds that give rise to the four legs of a dog and the arms and legs of a human. Only a few base pairs and their arrangement differ. The genes that code for hormones, neurotransmitters, and the other fundamental chemicals involved in fear, aggression, bonding, appeasement, joy, and grief are also conserved in varying degrees across mammalian and bird species.

Nature has been quite frugal with the raw materials that constitute so many different species, using a handful of chemicals such as cortisol, testosterone, oxytocin (sometimes called the cuddle chemical), and vasopressin; some bundles of nerves, cartilage, blood, and tissue; and five basic senses. The hormone oxytocin, which underlies bonding between mothers and infants, is essentially the same throughout the animal kingdom in species that rely on mother-infant bonds. Oxytocin and vasopressin are both highly conserved hormones.

A study that illustrates the power of genes was conducted on a mouselike creature known as a vole. The prairie vole is highly social and monogamous, lives in multigenerational families, and shares child rearing between males and females. The montane vole shuns company, lives alone, and is polygamous. The two types share 99 percent of their genes, but subtle differences in the remaining 1 percent appear to account for their very different social behaviors. One important subtlety lies in a gene that codes for cells in the brain called vasopressin receptors. Vasopressin is involved in regulating behavior, including sociability and monogamy, in male mammals. A single gene of the sociable male prairie vole contains a segment that causes these receptors to be distributed in a specific pattern not found in the montane vole. Tom Insel and Larry Young, of the Center for Behavioral Neuroscience at Emory University, created a strain of montane voles that carry the prairie vole gene and express its pattern of vasopressin receptors in the brain. With that single gene manipulation, the montane vole was transformed into a more social fellow with a preference for the company of females. After a small injection of the vasopressin hormone, the montane voles "began sitting side by side with females and grooming them," Insel says. Montane voles that were not engineered with the prairie vole gene ignored the females. "The ultimate goal is to understand the neural networks that govern complex social behaviors like monogamy," says Insel.

Just as the petals of the *Mucuna* blossom evolved to reflect bat radar and improve its chances of survival, hormones and neuro-chemicals have evolved in social species for a variety of functions that further their survival. In social species in which maternal care has evolved, the hormone oxytocin increases during pregnancy and kicks into high gear at birth. If a mother ignored her baby, the baby would have less of a chance of surviving. Oxytocin helps ensure that Mom is paying attention to that baby and promotes the mother-infant bond that we call maternal love. In other mammals, the genetic component of that bond evolved millions of years ago and has been conserved and passed along to all social species.

The genetic and physiological components of emotions are no different. They too have been conserved in mammals and birds and are very similar regardless of the species, underlying the ability of different species to predict each other's behavior. Many scientists argue that regardless of these similarities in emotional states, with-out a theory of mind an animal cannot be of aware its emotions. I will come back to this argument in the last chapter. For now, how-ever, there can be little doubt that humans and animals share much of their genetic code and are the products of tens of millions of years of trial and error called evolution.

Survival depends on an animal's ability to sense its environ-ment, communicate with members of its own species, and under-stand and predict the behavior of other species. Such pressures could have driven species toward a common natural language that would have benefitted all of them in their own selfish ways. Nat-ural selection would have favored signals that were understood to some extent by species sharing common environments. The hon-eyguide learned to communicate with humans and badgers and to cooperate to get at the honey and larvae inside the beehive. Ravens may lead predators to prey. The fact that mammals and birds express the basic emotions in similar ways, as we shall see in the next chapter, makes it easy for all of us to understand each other's motivational states and predict future behavior. Chirping birds and barking dogs may not sound very similar, but they are speaking the same natural language that we use every day.

Four

ONE LANGUAGE AND FEW WORDS

"... we respond to gestures with an extreme alertness and, one might almost say, in accordance with an elaborate and secret code that is written nowhere, known by none, and understood by all."

—EDWARD SAPIR, Anthropologist (1927)

 OMMUNICATING WITH ANIMALS can mean different ent things to different people. In recent years, telepathic communication, primarily with pets, has become a popular topic. Reputable people have told many otherwise inexplicable stories about animals and their owners communicating in this manner. Sadly, I am too feeble to converse mind-to-mind with animals, but to satisfy some friends, I tried communicating telepathically with the squirrel that lives in an oak tree outside the balcony of my aging sixth-floor Virginia apartment on the Potomac River. After all, squirrels are said to have the largest brain-to-body ratio of any animal.

The first time I tried such a thing, I started the communication with thoughts that would be unlikely to cause any cross-species misunderstanding. I asked the squirrel what it thought about the winter months, whether it had stored enough nuts, and how many children it had running around on the apartment grounds. After what seemed a sufficient pause, I detected no recognizable squirrel thoughts. The squirrel reacted to the telepathic inquiry as it normally does whenever I step out on the balcony—it scampered down the tree and hid on the other side of the trunk.

Perhaps I gave up too quickly, but this book is more concerned with the vocal, visual, and chemical signals that constitute the ancient natural language of animals. So I tried another approach that I had stumbled upon in a 1977 research paper published in *The American Naturalist* by Eugene Morton, titled *On the Occurrence and Significance of Motivation-Structural Rules in Some Bird and Mammal Sounds*. This is not a title that would normally capture a person's attention, but the first paragraph of Morton's paper was all it took to hook me. "Empirical data . . . showing that natural selection has resulted in the structural convergence of many animal sounds used in 'hostile' and 'friendly' contexts. Simply stated, birds and mammals use harsh, relatively low-frequency sounds when hostile and higher-frequency, more pure, tonelike sounds when frightened, appeasing, or approaching in a friendly manner. Thus there appears to be a general relationship between the physical structures of sounds and motivation underlying their use."

Morton is a scientist at the Smithsonian Conservation Research Center in Front Royal, Virginia, and a world-class expert on bird communication. After reading his paper, I returned to the squirrel communication experiment with a vocal approach. I waited one morning at my window until I saw the squirrel out on a limb and nonchalantly stepped out onto the balcony. As usual, the squirrel scampered away and hid out of view on the other side of the tree. I then attempted to "speak" squirrel à la Morton, which is to say I imitated what seemed to be squirrel sounds in an appeasing or approachable tone.

On Midway Island, I had been embarrassed when I had attempted to engage a group of juvenile albatrosses in a courting dance. They seemed frightened, and one tried to bite me as I was on my knees shaking my head and crooning the sky moo. I have never handled rejection well, so striking up a conversation with a squirrel was emotionally risky so soon after the albatross incident. Vocalizing like a squirrel requires quickly bringing the tongue away from the roof of the mouth, in the way you would coax a horse to trot, or express the sound tsk-tsk. Almost immediately the squirrel came from around the tree and stared at me. I was startled but I made a few more of the sounds. The squirrel raised its bushy tail, pointed its ears toward me, and began to approach, looking straight at me with genuine curiosity. Within a few seconds it had climbed onto a branch that stretched out nearest to my balcony. (For the record, I was not holding a piece of food, which would have been baiting the experiment.)

The squirrel cautiously investigated the situation, head fully extended, ears still pointed forward, tail raised about halfway and twitching only slightly. Faster tail twitching would be a visual signal conveying aggression or fear, or more likely a mixture of both. I made the sound again, taking care to sound as friendly as possible. Time seemed to stand still. The squirrel appeared to sniff. Its dark eyes were alert as it continued to gaze at me, now less than ten feet away. Then, having satisfied its curiosity and certain that I had nothing more of interest to offer, the squirrel scampered off. Morton might be onto something. Communication had occurred between man and squirrel, without provisioning. The experiment has been repeated many times since with other squirrels on the apartment grounds. Once a squirrel was bounding away in the opposite direction when I made the sound; it stopped in its tracks, turned, and came up to me. It stood on its hind legs and stared. A lot of people here do feed the squirrels, some of which appear to be habituated, as scientists say, to humans. I cannot communicate via thought waves, but I can amaze my friends with the ability to gather curious squirrels at my feet.

Morton is a real-life Dolittle, who understands the general

meanings of bird and mammal vocalizations and is able to communicate with them. He groups the types of calls influenced by his rule into three general categories: growls, barks, and whines. The following is the rule as it is described in Morton's paper:

1. Harsh, low-frequency sounds indicate that the sender is likely to attack if the receiver comes closer to the sender or remains at the same distance. In some species, the harsh end point is given only during the attack or when an attack is imminent.

2. Relatively tonal, high-frequency sounds indicate that the sender is submissive, will not be hostile if approached or if approaching, or is fearful.

3. Harsh sound quality, tonal quality, and sound frequency (pitch) interact such that: (a) the higher the frequency, the more fearful or friendly the sender; the lower the frequency, the more hostile; (b) the greater the sound's harshness, the greater the aggressive motivation; the more pure and tonelike, the more fearful or friendly, no matter what frequency range is used.

4. Sounds rising in frequency (no matter what the sound's quality) indicate a lower hostility or increasing appeasement or fear, but a motivational end point is not indicated. Sounds decreasing in frequency indicate an increasing hostile motivation.

5. A sound whose frequency rises and falls more or less equally or is constant but occurs midrange in the overall frequency range reflects a conflict of motivation over whether to approach or withdraw from a stimulus. It indicates a stimulus of "interest" has been received by the sender.

6. A species that is generally more aggressive to its own members will tend to have a harsher close-contact vocal repertoire, as opposed to a species that often joins or is joined in flocks, especially mixed-species groups. The latter will have a prevalence of high-frequency pure tonal sounds in its repertoire.

7. A species with greater complexity of social interactions will evolve sound signals containing a more complete range of sound qualities indicating more points along motivational (emotional) gradients and rapid changes in motivation than a species with less complex social interactions.

8. An individual uttering "alarm" sounds will be more likely to withdraw from the alarming stimulus the higher pitched its sounds, if the alarm system is graded. The alarm system will tend to be graded if kin are reliably nearby or if coordination of an escape as a group will reduce the chances of predation.

These rules of vocalization are so simple and universal that humans probably give them no more thought than they would to breathing. But consider the voices of the witches in *The Wizard of Oz*. Glinda the good witch has a sweet lilting voice, while the wicked witch of the west has a low, harsh voice. When my editor gets angry, she lowers her voice and says in a staccato manner, "If—you—ever—spend—that—much—on—assignment—again—you—will—be—writing—obituaries." When a person is friendly or appeasing, the tone of voice is higher and more legato, as in "How sweet of you," or "It is so nice to see you," or "What a pretty baby." The rule applies to both humans and animals.

After reading Morton's 1977 paper and looking up his academic affiliations, I discovered he was still actively involved in animal communication research and, incredibly, less than a two-hour drive from my home. We spoke by phone and he graciously agreed to meet. After I related the squirrel story (but not the telepathy part or the albatross incident), Morton explained that the sound I had imitated was a "bark," which is covered by number 5 above. Barks are a common vocalization used by many mammals and birds. The bark of a dog falls in between its growl and its higher-pitched whine. Any vocalization that falls into an animal's middle range of frequencies is its bark. The caw of a crow, the chirps of birds (but not their songs), and the meow of a cat are all "barks." According to Morton, mammals and birds use barks in a variety of contexts. They can generally be interpreted as: "I have discovered a stimulus of interest to me, and it is probably of interest to you," or "I am here and I am the same species as you are." When I made the squirrel sound, I did not say anything in particular, but the sound caught the squirrel's attention, and because it was understood as nonthreatening, the squirrel approached. Migratory birds often use barks to defend food resources or nonbreeding territories. Birds also employ barks as a mobbing call—a call to arms that signals other members of the species to rally against a predator such as an owl or a hawk. The birds cooperate to harass and chase the predator from their territory.

Morton led me outside his office in the Blue Ridge Mountains to a stand of trees where three or four dozen birds were busy foraging. He said that he could grab the attention of all the birds in the trees. This seemed to be a tough challenge because the birds were making quite a racket of cranky-sounding single-note chirps that might be translated as "Piss off—my branch." Birds spend a lot of time deliberately annoying each other. Morton brought the palm of his hand to his lips and made what sounded like a screechy birdcall—not songlike at all. The birds immediately stopped bickering and came from the interior of the tree to the outer branches to gaze at us for a few moments. Then, like the squirrel, they appeared to lose interest and resumed their bickering. What struck me most about the call Morton made was that it was generic—that is, not specific to the species in the tree. The birds were simply responding to the frequency and modulation of the call. This is noteworthy because it suggests that animals of widely varying species are able to understand each other's calls based on their degree of harshness or friendliness. They can understand each other's motivation or emotional state.

In their textbook *Animal Vocal Communication: A New Approach,* Morton and Donald Owings, a researcher into animal behavior at the University of California at Davis, compiled a table of bird and mammal vocal signals that fits the motivation-structural rules. These signals have been observed and reported in the scientific literature by numerous researchers. Morton and Owings list 30 species of birds and 29 species of mammals that vocalize aggression, hostility, appeasement, approachability, submission, and fear with basically the same patterns of frequencies. When low-frequency harsh sounds from various species are compared using spectrographic analysis, the sound structures look nearly identical.

Morton has had the good fortune to work and live in a beautiful preserve in Virginia with some of the most unusual species on the planet. The Smithsonian's Conservation Research Center was once the location of the world's most active breeding program for endangered species. Over two decades, Morton studied the calls of a wide variety of exotic animals and birds kept at the center and documented further consistency of the rules.

Morton has tested the motivation-structural rules on the Virginia opossum, the Tasmanian devil, and the rare maned wolf of Argentina, to name just a few. Each makes sounds with their own species-specific qualities, but the vocalizations of all the animals studied can be grouped into the same discrete categories of growls, barks, and whines. Within the category of whines, Morton has found that squeals and squeaks are primary types of appeasing or friendly calls. The rule applies primarily to the vocalizations animals make when they are in relatively close contact. Ordinarily it does not apply to the long calls that birds and mammals make when announcing their presence on established territories, such as bird-song or the songs of male humpback whales. But the low-frequency territorial roar of the Central American howler monkey may have evolved from the low-frequency aggressive calls it uses in close skirmishes with intruders.

One does not need to look any closer than the backyard to see and hear the rules at work. Purple martins are insect-eating migratory birds that readily accept the public housing units provided by generous people throughout the eastern and midwestern United States. When the birds arrive each year to take up residence in their multitiered apartments, the dominant martins will demand control of the penthouse suites at the top. Usually some early arrivals of lower-ranking social status have already moved into the empty penthouses. The dominant birds quickly set about dislodging the tenants. They put on aggressive displays, and the poor fellows being evicted from the prime roosts are unmistakably submissive. The aggressors puff up their feathers and make predictable low-frequency calls that sound like *"zwrack, zwrack zwrack"* when scolding the cowering interlopers. But when the same aggressors address their mates, they emit a call that sounds like *"sweet."*

Growls, barks, and whines do not qualify as a language by strict human linguistic standards, but they are highly effective means of communicating. Morton argues that the motivation-structural rules of animal communication are analogous to the universal rules of grammar found in human speech. Once a person becomes aware of the rules and general features of this natural language, it is relatively easy to distinguish between mating calls, territorial calls, aggressive encounters, and the more intimate conversations that take place among families and members of the same social groups. The adage "It's not what you say but how you say it" holds just as true for animals as it does for humans.

Even as I write, a house sparrow that has claimed the antiquated window air-conditioning unit in my bedroom as its turf, is chuttering aggressively at an intruder. Every year in late winter, male house sparrows begin claiming the air conditioners of my apartment building, a great location near the banks of the river, with lots of insects and plenty of food scraps to pilfer from the residents. The male sparrows have quite a few spats until they all are settled. After watching these sparrows come and go for the past seven years, I believe they attach some status to the apartments on the higher floors because those are always the first ones to be claimed. People attach status to higher floors, too, and we pay for an incredible view. Sparrows like the safety. The aggressive chutters of the birds turn to higher-pitched single-note chirps when the males begin calling for mates, and later on when they make territorial announcements.

The Canada geese that hang out most of the summer around the riverbank next to this building spend much of their time during the day foraging on the lawn. I now know when the geese are fighting without glancing out the window. I can hear loudly and clearly the characteristic harsh *gaeck, gaeck* they emit, usually when one of the geese wanders too close to another. They appear to dislike having their personal space invaded when dining. After dark, the sounds made by these geese soften considerably. On moonlit nights I can see their silhouettes as they float on the water near the shoreline, between the large round patches of duck grass. The geese make soft whimpers that sound like *kn* and *quais* when they are being friendly or appeasing. Their voices sound rather sweet and produce an emotional response that I would describe as soothing; they are certainly reassuring to the geese. At night they make these calls as a way to stay in contact with family members, like avian Waltons, who at the end of each episode of their television show would call to each other, "Good night, Mom. Good night, Dad. Good night, John Boy."

If a person hiking in the wilderness hears a low, guttural growl coming from the bushes, he or she knows intuitively to beware.

When dogs growl, they lower their voice, which takes on a harsh tone. The meaning is clear and innately understood across species. When dogs want something from us, they vocalize with a high-pitched whine that is friendly. The sound is effective and hard to resist.

John Fentress, founder of the Canadian Center for Wolf Research, took custody of a wild wolf cub when he was a student at Cambridge University in Britain. No one at the time (this was the 1960s) believed a wild wolf could be raised in captivity and assumed that it would eventually turn on its owner and attack. Fentress named the wolf Lupey and proved the assumption wrong: Lupey spent the rest of his life with Fentress. Fentress tells a story that illustrates Lupey's ability to communicate directly with people, adhering perfectly to the universal motivation-structural rules. At Cambridge, Fentress rented a room at a farm where Lupey could have plenty of room to run and play. Raising a wolf in town would have made too many humans uncomfortable. One day, Fentress walked into the yard to greet Lupey, who by this time was a couple of years old. As Fentress approached Lupey, the wolf snarled in a threatening low voice.

"I took a step back and then he wagged his tail," Fentress says. He approached again and Lupey snarled once more. When Fentress stepped back, Lupey wagged his tail and whined. With a little investigation, Fentress discovered what Lupey was trying to communicate. "Lupey was saying, Don't stand on my chicken," Fentress says. Lupey had received scraps of a chicken and had buried them exactly where Fentress had been standing. Says Fentress, "When I stood back, he was friendly again. If animals are not happy with what you are doing, they will convey it without killing you."

Fentress has spent years studying patterns of communication in wolves. When wolves greet each other, they make high-frequency whines similar to the sound of a dog whistle. In aggressive contexts, they growl deeply as a threat. If a wolf falls silent during an aggressive encounter and its tail is in an upward position, it is threatening to attack. Howls are territorial calls and also prompt members of a group to join each other. Wolves display a broad range of emotions that are expressed according to Morton's motivation-structural rules, but motivation carries the additional implication that the receiver can predict the future behavior of the sender based on the tone of its voice.

After the tragedy of September 11, 2001, I spent four months covering the war in Afghanistan and its aftermath. For the first month I lived in a tent near the headquarters of the Northern Alliance outside a dusty, barren village called Hoja Bahauddin. Because of the time difference between Afghanistan and Washington, D.C., I spent a lot of late hours talking with editors on a satellite phone outside my tent. The area was far enough from the front line to be safe from the Taliban's artillery at night, but it was not protected from the packs of wild dogs that roam all over that devastated country. They usually came around after 10 p.m., when a curfew took most people off the dirt streets. The wild dogs owned the night, and the packs howled at each other until near dawn. Some of the dogs looked as if they weighed 70 pounds, and even the grizzled Northern Alliance soldiers seemed afraid of them. The Afghan people do not keep dogs as pets and find the idea downright odd. Their dogs are only for dogfights and gambling.

Sometimes, when I was standing out in the Northern Alliance's compound, talking on the sat phone, the dogs would gather about 30 feet away from me and begin low, threatening growls. The dogs made regular nighttime rounds to raid the soldiers' garbage, which frankly did not offer much. I suspected that the dogs viewed me as standing in their way. The growls tended to produce an uncomfortable tension between us and made it difficult for me to focus on my conversations with Washington. It was probably a dumb idea, but one night I decided to apply what I knew about the motivation-structural rules, growl back, and defend my new turf. I squared my shoulders and spread my arms to make myself look bigger, which was not hard considering the thick layers of clothing, stocking hat, and scarf needed to stay warm. In the lowest voice I could muster, I growled loudly like some movie monster at the small pack of hounds while my befuddled editor waited on the other end of the line. The dogs seemed to be surprised. They did not attack, nor did they challenge me on any other night. They still made their rounds, but we seemed to have reached an understanding to leave each other alone. Sometimes, though, I would wake up and find them sniffing at my tent. I could see their distorted shadows reflected on the tent walls by the moonlight. They could easily have torn through the thin fabric, but they never did.

Later on, after the Taliban fled, I moved to a house in Kabul, set up a bureau, and rented rooms to other reporters. A reporter for the *Kansas City Star,* who was staying in our rather luxurious home, rescued a puppy from a dogfight. Puppies were used to bait the fighting dogs and get them into a killing frenzy. Malcolm pulled the pup out of a trash can and brought it home, much to the chagrin of the Afghan staff, who held the dog in the same regard as one might a wild rat. Named Maggot, the puppy was a little wild animal with aggressive instincts. When he attacked a foot or an arm, he would growl viciously and shake his head. Of course he was also barely eight weeks old and wanted to play constantly, which he would signal by whining, extending his front paws, and bowing with his tail in the air—a classic canine play = bow signal, used by dogs, coyotes, and wolves. Maggot liked the reporters in the house, but if any of us approached him when he was eating, he would growl as low as he could and snap, "Don't stand on my chicken." Maggot was with us for about two months, and then a plan was hatched to forge papers and smuggle him into Germany with another *USA Today* reporter, who was stationed in Berlin. The plan worked and Maggot was flown out of Afghanistan on a United Nations charter flight, placed with a German family, and sent to dog school. Germans take their dogs very seriously. As a footnote, by the time Maggot left Kabul for his new life, the Afghan staff had grown to like him and played with him as much as we reporters did.

Low-frequency sounds originated as a reflection of aggression and dominance in the distant evolutionary past of vocal vertebrates—in reptiles and amphibians, which preceded the appearance of mammals and birds. To examine the evolution of a species or trait or some aspect of behavior that leaves no fossils, scientists can study living animals that can trace their lineages to older species. Unlike mammals and birds, amphibians continue to grow throughout their lives. In nature, bigger is usually better. With bullfrogs, for instance, large size is an accurate indicator of age and survivability. The bullfrog is also a classic illustration of the origin of low-frequency vocal signals.

The Ozark Mountains, where I grew up, are located in a temperate region along the Missouri and Arkansas border and are one of the oldest ranges on earth. They contain an unusual diversity of animals, including an orange-throated lizard that runs on its hind legs like one of the small raptors in *Jurassic Park*. In the spring and summer, the region looks a lot like a jungle. My grandparents lived on a remote hilltop and had a pond near their house that was thick with American bullfrogs, the largest North American frog. Occasionally these frogs prey on small birds and snakes. It is a rare treat to see a bullfrog catch a bird. On summer evenings, male bullfrogs would gather at the water's edge and begin serenading the females with their deep, vibrant chorus. I would crawl slowly and quietly to the top of the red clay pond bank, being careful not to disturb the frogs, and listen to the concert. The females floated about 15 or 20 feet out from the bank, with their heads and ears raised above the surface, and their legs and flippers dangling behind them, like little scuba divers. When the females hear a deep and strong croak that they like, they swim to the male and lay their eggs for him to fertilize. Females prefer mates that have the deepest croaks, and those frogs naturally have the biggest bodies.

The evening bullfrog serenade is serious business. Before the concert, the males push and lunge at each other and wrestle with their forearms for the best positions at the water's edge. Larger males usually win conflicts with smaller males for favored territories on the pond's bank, but fighting can be risky, especially for the smaller bullfrogs. Somewhere in the evolutionary past, animals learned that a low, deep voice was a good indicator of size and strength. With the rise of communication as a substitute for violence, animals realized that if an opponent had a deeper voice, it paid to back off rather than risk injury. Communicating rather than fighting was favored by natural selection, and the low-frequency sounds emitted by larger animals came to be linked with successful aggressive encounters and dominance and thus became the basis of growls. Since the bullfrogs and other animals that emitted the lowest sounds were bigger, they had the bonus of being favored by females, so that the force of sexual selection also came into play. Frogs that could expand their throats could get a deeper, more resonating sound. Eventually, they developed the balloonlike sacs that enhance the resonance and low frequency of the croak. This trait, in turn, made some bullfrogs appear larger and more intimidating to competing males, in front of which they expand their throat sacs to appear as large as possible.

Looking big is universally important to mammals, including humans, and birds during aggressive conflicts. When I made myself look big to the hounds in Afghanistan, I was speaking their nonverbal language as well as my own. I spread my arms, squared my shoulders, and growled to say, "I'm bigger and scarier than you, so leave me alone." Birds will partially expand their wings, or sometimes fold them back to appear larger. When they arrive at a feeder in the backyard, they will raise their wings and sometimes lunge at each other as they jockey for the best position at the trough. I often put breadcrumbs out on the ledge of my balcony for the house sparrows during the late winter. Like most birds, sparrows are very aggressive with each other. When conflicts arise, their feathers stand up a bit as they chutter harshly. The older and larger male house sparrows, which possess a dark badge of dominance on their breast, usually win and get first pickings of the crumbs.

Among mammals, one of the primary ways of making oneself look bigger to an opponent is to raise the hair on the body. This cannot be done consciously. Piloerection is an involuntary physiological response to fear that is common in both mammals and birds. It happens reflexively in us humans, too. If a person walks into a darkened alley and senses someone walking up from behind, the fine hairs on the back of the neck and arms will stand up without any conscious will involved. When chimpanzees are threatened or frightened, the hair all over their bodies stands on end. During the filming of Jane Goodall's IMAX film on the Gombe chimps, one of the young chimpanzees, a toddler, became extremely curious about the large IMAX camera. He wanted to race over and explore it, but the camera was a new and frightening object. Finally, with his coarse dark hair raised all over his little body, the chimpanzee ran toward the camera, touched it, screamed, and ran away. (He also had an erection, which is another involuntary response to fear caused by dilated blood vessels.)

Animals do not want to fight if it is not necessary. The exceptions appear to be humans, chimpanzees, and dolphins, all of which are known to attack and kill members of their own species without apparent provocation. To win a threatening display, whichever animal carries on with its bluff the longest declares victory. Sometimes the smaller of two opponents will win by growling and displaying for a longer period. In this case, the littler guy is saying that he is highly motivated to win, probably because he has more to lose. If the bigger opponent is not as motivated, he will usually back down rather than risk being hurt.

In any aggressive conflict, a signal is needed to communicate "I give up." Natural selection has provided a clear, universal signal for defeat. In *The Expression of Emotions in Man and Animals,* Darwin describes a phenomenon he calls the principle of antithesis. At the heart of the motivation-structural rules, this principle applies to vocal signals and visual displays. If the aggressor is using a low-frequency harsh sound and making its body look bigger, then the simplest, clearest way to express "uncle" is to do the exact opposite. To submit, the animal whimpers or screams with a high-frequency sound and cowers. The animal makes itself sound as meek and appear as small as possible. The sounds and physical postures associated with aggression and submission, however they are expressed by a species, are always opposites. Ambiguity causes misunderstandings that could lead to injury. If a person raises his fist to strike, the reflexive response is to turn away and become as small a target as possible. The movements associated with submission arise from the natural responses to aggression. A submissive dog bows its back downward, raises its head toward the aggressor, tucks its tail between its legs, and whines. Birds do the same thing. Like the purple martins, the aggressor fluffs its feathers, leans forward, and raises its wings, while the submissive one bows its head with its wings tight against the body and tail feathers in the air. The aggressor sounds raspy while the submissive peeps.

The female bullfrogs' preference for deep voices in males appears to carry across many species' boundaries. Sarah Collins of the Institute of Evolutionary & Ecological Sciences in the Netherlands conducted a study of the effect of men's voices on women, titled *Men's Voices and Women's Choices.* To learn what judgments women make about the vocal characteristics of men, Collins recorded the voices of 34 males uttering five vowel sounds and analyzed them for various features, such as harmonic frequencies. The males were measured—height and weight and hip and shoulder width—and were also asked to describe whether they had chest hair and how old they were. Collins played the voices of the men to a group of women who were asked to guess about the men's physical attractiveness, age, weight, height, degree of muscularity, and whether they had hairy chests. The women judged that the voices lower in frequency and more closely spaced harmonically belonged to men who were more attractive, older and heavier, and more likely to have a hairy chest and to be muscular. They were wrong on every count except weight. Nonetheless, even though the women guessed incorrectly, the study reveals the assumptions some women make. One study does not prove the theory, but these findings bear a striking resemblance to the preferences of females in other vertebrate species, which have been closely documented.

Stanford Gregory, Jr., and Timothy Gallagher, sociologists at Kent State University, have conducted a series of intriguing studies on pitch and dominance effects in human voices. One of the first studies, conducted by Gregory in the mid-1980s, found that people tend to adapt the frequencies of their voices to each other during a conversation. As the studies were refined, Gregory discovered that the pitch of a person's voice was the factor people were responding to when interacting. Specifically, frequencies below 500 hertz are perceived as pitch. The dominant, low-frequency roar of a howler monkey is at 360 Hz. The upper and lower limits of human hearing are about 20 Hz to 20,000 Hz, but the most sensitive range is 1,000 to 4,000 Hz. Gregory called the band below 500 Hz the fundamental frequency of phonation and believes that it reveals unconscious information about a person's level of dominance or perceived social status.

Studies by other scientists, conducted in the early 1990s, found that people of lower social status accommodate their nonverbal vocal patterns to people of higher status when interacting. The lower-status individual might begin matching the cadence and pitch of his boss's voice when called into her office for a review or even a casual conversation. The higher-status person does not accommodate her voice much at all. This phenomenon became known as communication accommodation. In 1996, Gregory led a study involving 25 celebrity guests appearing on the *Larry King Live* television talk show. The results showed that King accommodated his vocal patterns to guests of higher social status. Guests of lower social status accommodated their voices to King.

Research on human nonverbal communication has revealed valuable information. Paul Eckman, a professor of psychology at the University of California at San Francisco, conducted studies in the 1960s that demonstrated that nonverbal communication involves vocal pitch, body language, facial expressions, and other factors. According to Gregory, individuals seem to have less conscious control of their nonverbal communications, which tend to reflect the true feelings or message contained in a conversation—more so than human speech alone. Most people have probably had the experience that something is not quite right during a conversation. The other person may be saying the right things, but the message coming across is different—tone and body language do not match the informational content of the words. Nostrils flare, pupils constrict or dilate, the face flushes, eyes look away. You can be a terrific natural lie detector if you trust your instincts. The ability to assess nonverbal cues such as these is a product of natural selection and first evolved in animals for sensing danger and deception.

The human brain is similar in its operation to a computer with parallel processors. Nonverbal messages contained in pitch and in body language are being produced below the conscious level while you are consciously speaking and choosing your words. The receiver is interpreting these messages unconsciously, too. This vocal encoding and decoding, which occurs in the grunts and screams of many primates, is usually associated with threats and dominance displays. Studies by Judee Burgoon, professor of communications at the University of Arizona at Tucson, have concluded that the human voice is more influential than visual signals at determining dominance because nonverbal vocal signals such as groans, growls, and sighs are innately produced and innately recog-

nized. A number of studies suggest that people who vary their intonation more than others when speaking are unconsciously regarded
as more dominant. Gregory's Larry King study supported that idea:
the socially dominant celebrities showed the greatest variation in
intonation.

Julian Keenan, assistant professor of psychology and director of
the cognitive neuroimaging laboratory at Montclair State University in New Jersey, studies differences in the right and left hemispheres of the human brain. Using functional magnetic resonance
imaging (fMRI), Keenan has confirmed that the right hemisphere
is "highly involved in nonverbal communication, such as hand gestures and intonation, and appears to be important for general facial
recognition, humor and deception detection." This had been
assumed, but the fMRI studies proved that the right hemisphere
"lights up" on the brain imaging scan during nonverbal communication. People who have suffered a stroke or other damage to certain areas of the right hemisphere lose their facility for this type of
communication and speak with a monotone voice. The right and
left hemispheres of the brain function as parallel processors, allowing humans to assess nonverbal signals at the same time they are
using language.

Gregory and Gallagher put all the theories about pitch, voice accommodation, and intonation and dominance to the test in a remarkable study that analyzed the nonverbal communication of
U.S. presidential candidates in every presidential debate conducted
since 1960. (Johnson held no debate in 1964, and Nixon avoided
debates in 1968 and 1972. His only debate occurred in 1960 and became infamous for his profuse sweating.) The pitch of each candidate in the debates was examined with spectral analysis, which has
the ability to filter out all bands except the fundamental frequency
of phonation. According to Gregory, the point of the study was to
determine whether the dominant partner in each presidential debate could be distinguished from the subordinate partner based on

pitch and voice accommodation and whether the dominant partner received the highest percentage of the popular vote in each election. The results found that the candidates' pitch predicted the outcome of the eight elections that were analyzed. In every case the candidates' nonverbal vocalizations revealed which candidate was dominant and which candidate accommodated the other. Nonverbal vocalizations predicted 100 percent of the time which candidate would win the popular vote. The lower and thus more dominant pitch of the winning candidate's voice was the key. Notably, the only candidate to lose a presidential election who was both dominant and the winner of the popular vote was Al Gore.

Gregory and Gallagher concluded: "Perhaps our finding—that analysis of a near-insignificant low-frequency . . . sound can predict the results of important contests for the most powerful political position in the world—suggests that anthropologist Ray Birdwhistell, one of the earliest and most eminent researchers of nonverbal communication, may not be far off the mark" in estimating that 65 percent of the social meaning in human interactions is conveyed by nonverbal cues. Others have put the estimate as high as 90 percent.

Natural selection has produced a universal language of nonverbal signals that is spoken and understood by animals and humans alike. To discover that the nation's selection of presidents is (usually) based on the same signals that mammals have been assessing to determine the outcome of aggressive interactions is powerful evidence that we speak the natural language of animals and rely on it more than information to make important decisions.

Just as low-frequency sounds signal aggression and dominance, the higher-frequency sounds have evolved to convey appeasement and approachability, but these sounds follow an entirely different physiological and motivational path. According to Eugene Morton, they are linked to the powerful innate response of adults to infant vocalizations, which is to provide care. Few differences can be found between the peeping of hatchlings, the high-frequency meows of kittens, and the cries of a human baby. All three sounds have been favored by natural selection and invoke an ancient response. The expression *sweet talk* connotes a deliberate, effective manipulation of the voice used by people to get something they want—true to the definition of communication as manipulation of behavior. It is the tone of the voice that gets the response rather than the words. When one of my daughters wants something, usually money, the endgame sounds like, "Please, please, please, please, please, please?" It is difficult to resist. Their begging calls sound a lot like the cheeping of little birds, which usually induces feeding by the parents. After a hatchling is able to see, it increases the pitch and rate of its begging calls when the parent arrives with dinner, competing to be fed first.

Further evidence that human nonverbal expressions are part of a natural language shared by all animals is found in the interaction between a mother and her baby. Most mothers seem quite naturally equipped to communicate nonverbally with their babies, and the infants respond as if they understand. Nonverbal baby talk includes melodic scales of intonation that rise and fall and garner wide-eyed expressions and smiles from the infant. Vocalizations are exaggerated and usually have a slow cadence and overall high pitch that falls perfectly in line with the expression of appeasement and approachability in animals. People make similar sounds with their pets, and people who have not had children need no coaching to speak in the lilting high frequency of baby talk. The gentle contact calls of the geese on the Potomac River, the purring of a cat with its kittens, and the soft grunts of chimpanzee mothers with their offspring follow the same pattern.

With human infants, parents must say no at some point before the babies learn to speak. An infant does not understand *Webster's* definition of *no,* but it clearly understands the tone of the voice. When a mother must protect her infant from picking up something dangerous, she quite naturally lowers her voice, and her normally legato tone becomes more staccato. This always gets the baby's attention. I have a vivid memory of being on my grandparents' porch at a very young age, when I saw an interesting little creature crawling along—it was brown with little claws in the front and a tail that curved up over its back. As I reached for the scorpion, a number of adults shouted at me fearfully in unison, which viscerally got my attention.

The psychologist Ann Fernald published a paper on the subject in 1992 in *The Adapted Mind, Evolutionary Psychology and the Generation of Culture,* with the long title "Human Maternal Vocalizations to Infants as Biologically Relevant Signals: An Evolutionary Perspective." Fernald observed that the exaggerated, stereotyped vocal patterns in mothers' speech are found in all human cultures and indeed fit the same pattern seen in nature and in intonations used among adults. Linguists have usually argued that the purpose of infant-directed speech (mother cooing to the baby and exaggerating sounds) is to teach language to the infant. But Fernald took a different point of view in her paper: "The fact that mothers quite consistently simplify their speech and use exaggerated intonation even when interacting with newborns, long before language learning is a central developmental issue, suggests that the modifications in infant-directed speech serve pre-linguistic functions as well."

This means mothers are communicating nonverbally to stimulate the emotions. When Mom smiles, her baby smiles. If Mom makes a sad face, her baby mirrors her expression. Infants of depressed mothers can show delayed emotional development because they do not receive normal emotional stimulation through facial expressions and intonation. Mothers typically exaggerate their facial expressions when vocalizing, which Fernald said reveals to the infant the feelings and intentions of others. Only toward the end of the first year does the prosody of the mother's voice begin to highlight words to help the infant identify linguistic units in a sentence or stream of speech. Other research reviewed in Fernald's paper demonstrates that moderately intense sounds cause the heart rate to slow down and make the infant orient to the source of the sound. Higher-intensity signals increase the heart rate, and the infant becomes defensive. Signals with a gradual rise and fall in their intensity cause the infant to open its eyes wide and look at the source. Abrupt or staccato sounds result in the infant's closing its eyes and withdrawing.

Fernald's research found that approval vocalizations in English, French, German, and Italian are higher in frequency, with a prominent rise and fall contour. "Prohibition vocalizations" like "Cut that out right now!" are typically lower in frequency, shorter, more abrupt, and more intense—more like growls. Comforting vocalizations tend to be held longer and are less intense. Other studies have documented that adults can accurately assess emotions from the sound of another adult's voice. The patterns of sounds that express happiness and enjoyment and those that express irritability and anger fit perfectly with the patterns of the motivation-structural rules of mammals and birds.

Fernald took her research a step further and tested the responsiveness of five-month-old infants to negative and positive facial expressions. Some infants reacted and some did not. She then tested the infants' responsiveness to positive and negative speech directed at them and compared their reactions when the adult used a tone of voice more typical of adult conversations. The infants responded strongly to the tone of voice we use for baby talk but did not respond to the adult tone of voice. The infants from English-speaking families also responded to baby talk that was in spoken in German, Italian, and Japanese.

Fernald's study concluded that the mother's intonation has a direct physiological ability to induce emotion in the infant and that the mother's speech does not teach the baby actual language in the early months of development. Fernald compared these physiological effects with those created by mother-infant bonds and forms of communication—both tactile and auditory—in other species. She found striking similarities in the use of tone and touch between mothers and infants in nonhuman primates. Mom's voice and its effect on her baby's emotional arousal play a significant role in the infant's overall healthy development. The importance of a mother's voice is probably underappreciated, since humans tend to focus on words and their informational content rather than on the power of voices to express and influence emotions. Fernald states: "Given our reliance on linguistic symbols to decode meanings in speech, it is perhaps difficult for us to appreciate just how important it is to be able to communicate through emotional signals. The predisposition to be moved by as well as to interpret the emotions of others and the ability to discern the intentions and motivations of others through expressions of the voice and face are remarkable evolutionary advances in communicative potential among the higher primates. Human mothers intuitively and skillfully use melodic vocalizations to soothe, to arouse, to warn and to delight their infants, and to share and communicate emotion. While the acquisition of language will eventually give the child access to other minds that are immeasurably more powerful and intricate than that of other primates, human symbolic communication builds on our primate legacy, a foundation of affective communication established in the preverbal period."

According to David Givens, director of the Center for Nonverbal Studies in Spokane, Washington, "A significant number of voice qualities are universal across all human cultures, though they are also subject to cultural modification and shaping. Around the world . . . adults use higher pitched voices to speak to infants and young children. The softer pitch is innately friendly, and suggests a nonaggressive, nonhostile pose. With each other, men and women use higher pitched voices in greetings and in courtship, to show harmlessness and to invite physical closeness. In almost every language, speakers use a rising intonation to ask a question. The higher register appeases the request for information, and is often accompanied by diffident palm-up gestures and by submissive shoulder shrugs. The human brain is programmed to respond with specific emotions to specific vocal sounds."

The late anthropologist Gregory Bateson, author of *Mind and Nature*, argued that nonverbal communication is as important as language or more so in human interaction. He rejected the linguists' and traditional Cartesian thinkers' impression that language was replacing nonverbal expressions in the evolution of the human. "If . . . verbal language were in any sense an evolutionary replacement of communication by means of kinesics and paralanguage, we would expect the old, predominantly iconic systems to have undergone conspicuous decay. Clearly they have not. Rather, the kinesics of men have become richer and more complex, and paralanguage has blossomed side by side with the evolution of verbal language."

Kinesics refers to the field of study developed in the early twentieth century that investigates human bodily movements, facial expressions, and voice tones as methods of communication or accompaniments to speech. *Paralanguage* refers to the use of these nonverbal signals.

Numerous studies have found differences in the sexes' expression and perception of nonverbal signals in humans. Women, who are traditionally regarded as more emotionally expressive than men, are in fact better able to read nonverbal signals than men, and they are more likely to make eye contact and study faces. Women smile more often than men and are more attracted to other people who smile. Women also touch other people more often, except in the early phases of dating or when meeting prospective dates for the first time. Some studies have suggested that men's and women's hand gestures are different, but the reports are mixed. Other studies suggest that women use gestures more often in the presence of men.

Either way, hand gestures represent an important form of nonverbal communication unique to people and other primates. We can use our hands as tools to express ourselves. Every species uses what it has at its disposal, whether that is a tail, a fin, or flexible ears. The great apes—humans, chimpanzees, gorillas, bonobos, and orangutans—have evolved a rich vocabulary of hand gestures to add to their limited repertoire of vocal sounds. Chimpanzees' gestures seem to have semantic meanings. Like humans, they point to draw the attention of another to a particular object or direction, and they hold out their hand palm up to say "Give me some." Chimpanzees use the begging gesture to encourage a troop member to share fruit or meat after a kill. Universally understood, this primate gesture has been adopted by beggars all over the world.

Chimpanzees have been shown to demonstrate numerous examples of cultural transmission of behavior. Among these are hand gestures that may represent chimpanzee dialects. Chimpanzees are less vocal than people and rely more on gestures to communicate. That their gestures have dialects suggests to some scientists that young chimpanzees are learning gestures and their meanings by observing adults, much as human infants must learn speech from listening to adults.

Wild African chimpanzees solicit grooming with gestures that differ slightly from location to location. Grooming is a highly social behavior during which one chimpanzee sits patiently while another picks through its coat to remove parasites. It is a bit like humans trading back rubs. Grooming has a calming effect. Jane Goodall has compared grooming to the exchange of pleasant but trivial conversation between adult humans as a means of maintaining social bonds or trying to win points with someone of higher status. Mother chimpanzees often groom their children; daughters more often than sons groom their mothers; females groom dominant males with the intention of reducing the chances of being slapped in the future and also to solicit sex; and males groom each other to maintain social order, and often with an "I scratch your back, you scratch my back" intent. A gesture called the grooming handclasp was first identified in the 1970s.

"The Kasakela chimpanzees of Gombe raised one arm straight into the air above their head to solicit grooming, while the K community of the Mahale Mountains approached each other at the beginning of a grooming session and each of the participants simultaneously raised an arm over their heads and then grasped each other's hands. Though similar to the Gombe pattern, this is a dialectical variant of that pattern," Roger Fouts and colleagues reported. In their review of studies conducted by numerous scientists, they also found variations of the grooming handclasp between two neighboring populations in the Mahale region. If this behavior was genetic, it would not be expected to vary between such close neighbors, which sometimes mate with each other. "The K community prefers palm-to-palm clasps while the M community prefers a variation of this pattern, which is generally wrist-to-wrist." The reason for the wrist-to-wrist variation may have something to do with the M group's distinctive way of respecting age and dominance. In dominant-subordinate interactions, the subordinate supports the arm of the dominant individual. The differences may also serve as a group badge of identity. Members of the K group can easily recognize members of the M group based on their style of handclasp.

In the rancorous debates over the origin of language, experts in nonverbal communication argue that early humans most likely relied on hand gestures similar to that of chimpanzees. Hand gestures can convey a great deal of information, and their use is found in all human cultures and in blind infants and young children. Because primates gesture so often but are unable to form sounds the way humans do, psychologists made the decision in the 1920s to teach apes sign language rather than spoken language, with some success. The resulting studies do not reveal how great apes communicate in the wild, but they do show that all of the great apes appear to have an innate understanding of the rules of grammar and syntax. How or whether they apply those rules to communicate through gestures is still being investigated.

Nonverbal hand gestures in humans and nonhuman primates are not to be confused with the signed language of the deaf. Although words are represented by hand movements rather than sound, signed language is a true language in the strict linguistic sense. In the book *Train Go Sorry: Inside a Deaf World,* Leah Hager Cohen points out that signed language has an expressive style. Children tend to pick up their parents' style much as they would accents used with spoken language.

From the gestures of primates, to the growls of wolves, to the voices of presidents, nonverbal communication is the "one language with few words" of humans and animals. Even though people have developed a remarkable spoken and written language for transmitting information, our nonverbal communication remains as important to our lives as it is to the lives of animals. The expressions of the emotions have been driven down a single evolutionary path by the forces of natural selection and sexual selection so that the earth's creatures can truly understand each other. The services of Dr. Dolittle have never been required. Humans and animals speak the same language and are obsessed with the same topics of conversation, including sex and real estate. Now it is time to eavesdrop on some of those conversations.

Five

THE CHEMISTRY OF LOVE

T HE VERY FIRST conversation to take place on the young earth some 3.8 billion years ago would have been conducted in a chemical language between two bacteria—a simple exchange that basically indicated "I'm here," "Me too," and "Divide and conquer." Where bacteria came from and how they came to be are central questions in discussions about the origin of life on earth. Perhaps the organisms were hitchhikers, residents on an asteroid that slammed into the planet's new ocean. Maybe the bacteria had always been here, living in the earth's molten core— certainly if any group of organisms could thrive in hell, it would be bacteria. The late Carl Sagan believed that bacteria are ubiquitous, living in all manner of environments throughout the universe. The late Stephen Jay Gould described them in his book *Full House* as the most successful of all life-forms to have evolved.

Whatever the origin of bacteria, they could not have survived and built the first community without a way to communicate. Chemical signaling occurs when a cell releases a dollop of some substance through a channel in its membrane that is then sensed by another cell's receptor. To be true to the sender-receiver definition of communication, the signal must cause some change to occur between the two organisms. Remarkably, the first communication to arise between the first two cells on this planet so long ago would one day become the common medium for the chemistry of love.

The first community of single-celled life-forms might have been something called a biofilm. Scientists have only recently discovered the highly evolved, intricate communication systems of these bacterial dynasties, which are capable of forming in virtually any type of ecosystem. They thrive as yellow mats on the surface of rocks inside and around the superheated geysers of Yellowstone National Park. They are found in massive numbers inside bizarre natural towers called black smokers on the ocean floor. These deep-ocean, "thermophilic" bacteria do not require sunlight to survive, and they do not eat organic matter. They dine on carbon dioxide, methane, ammonia, sulfur, and other inorganic compounds. Biofilms also grow in large mats in swamps and can be found inside glacial ice. They might have formed colonies in the oceans of Io, one of Jupiter's many moons, and inside the red soils of Mars. Biofilms also make a person's teeth slippery and cause ear infections, salmonella poisoning, and the deterioration of the hulls of ships.

Bacteria have developed many different chemical languages, but biofilms are often made up of multiple species that communicate through a universal language of their own. The common language of bacteria is the oldest language on earth. Bonnie Bassler, assistant professor of molecular biology at Princeton University, has broken at least part of the code of this language. She and her colleagues have discovered that it appears to be based on a single gene, LuxS, which is shared by different species of bacteria. The gene is responsible for manufacturing a chemical, a pheromone to be exact, that allows the bacteria to sense each other's presence and determine how many are in the neighborhood. Microbiologists call this quorum sensing. The pheromone-based quorum-sensing signal helps coordinate the gene expression of all of the community's organisms, which allows the population to act in unison as a single superorganism.

Bassler made her discovery by studying two species of ancient bioluminescent bacteria, *Vibrio harveyi* and *Vibrio fischeri,* which are able to emit a blue glow. These bacteria use the LuxS gene to generate the same pheromone signal to survey the numbers of their local population. Each member of the community converses by means of the signal—taking roll call, dividing, and adding new members until the population's concentration reaches a critical mass. In that moment, when they have reached a quorum, the community cries in unison, "Let there be light." *V. harveyi* swim freely in the ocean and colonize fish species and plankton. They glow when disturbed, so the sudden flashes they produce probably function to frighten predators. Both these bacterial species make use of the same gene and pheromone to communicate with each other. The same type of quorum sensing has been incorporated into the genomes of more onerous organisms, which develop into the slimy, impenetrable biofilms that resist antibiotics, spread cholera, and turn a package of hamburger lethal.

The harmless *V. fischeri* colonize a chamber that circulates seawater in the bobtail squid, which makes its home beneath the sand in the shallow coastal waters of the Pacific Ocean. The squid is commonly found around the Hawaiian Islands, where it hunts at night. Its predators swim at lower depths and scan the waters above for the squid's moonlit silhouette. The bacteria multiply throughout the day inside the squid. By nightfall, through quorum sensing, they produce light in the squid's light organ. A tiny sensor, which assesses the amount of moonlight, determines how bright the squid's body should glow. The bacteria produce enough light to eliminate the squid's silhouette, making the squid invisible to predators below.

Bassler's studies of these species have turned the field of microbiology on its head. Until recently, scientists did not believe that bacteria could communicate, but this kind of animal talk is now one of the hottest fields of microbial research, promising to unlock secrets for conquering microbial diseases.

Bacteria had the earth to themselves for more than 3 billion years, plenty of time to develop a sophisticated level of chemical signaling through pheromones and their corresponding receptors. Before the earth had an atmosphere that contained oxygen, anaerobic bacteria would have been the only type of life capable of surviving. The "endosymbiotic theory" suggests that at some point, maybe around 2.8 billion years ago, as the atmosphere began to accumulate oxygen, bacteria that had evolved an ability to synthesize oxygen would have gained a tremendous survival advantage over the planet's many anaerobic populations, which would have been killed by oxygen. The theory posits that an anaerobic amoebic bacteria, such as blue-green algae, tried to eat one of the new, aerobic varieties. Unable to digest it, the anaerobe found itself with this new thing inside that possessed the ability to use oxygen to make energy, and a symbiotic relationship was formed that would have helped the cell adapt and survive. This favorability led to the evolution of multicelled organisms and gave rise to mitochondria, which are essentially bacterial cells that reside in the cells of all multicellular organisms today. Mitochondria have become the energy factories of cells in animals, including humans. The original relationship between the aerobic and the anaerobic bacteria would not have lasted without the occurrence of chemical communication between the two cells.

About 500 million years ago vertebrates appear in the earth's fossil record rather suddenly—in evolutionary terms that is, whereby "sudden" involves hundreds of thousands of years—in what is known as the Cambrian explosion, a proliferation of new types of creatures found in rock strata of the Cambrian geologic period. The chemical signaling of bacteria provided the foundation for cell-to-cell signaling and enabled the evolution of more complex organisms, including, ultimately, the human body and brain. Pheromones, the remarkably powerful main catalysts for the chemistry of love, are evolutionary gifts from the ancient bacteria.

Bacteria offer many great services to living organisms and in turn derive benefit from using their hosts as sources of food. The bacteria that commonly reside in the human gut get all the nutrients they need and in turn aid our digestion. Anyone who has taken antibiotics that wiped out bad bacterial fauna from the intestines along with the good knows how important they are. Most lifeforms on earth are merely hosts and sources for bacterial opportunities. Some scientists joke that bacteria have facilitated the evolution of higher organisms merely for selfish purposes. Bacteria certainly make good use of the human body and have managed to insinuate their DNA—the mitochondria of cells—into our own.

Bioluminescence was eventually co-opted for communication by higher organisms, such as jellyfish and anglerfish, the strange little sea monsters mentioned in chapter one. It is the most common form of communication in the oceans but is rare among terrestrial species. One of the very few insects to have adopted this chemical language of light is the firefly or lightning bug. The firefly is among the most amorous of creatures, devoting much its energy during its brief lifetime to producing flashes of light to attract a mate. A common feature of the backyard landscape in the midwestern, mid-Atlantic, and southeastern parts of the United States, fireflies are actually a type of beetle. Scientists have counted 136 different species around the world. A summer ritual for children where I grew up in the Ozarks was to arm themselves with glass mason jars, with breathing holes punched into the metal lids, and catch as many lightning bugs as possible. No harm was intended toward the poor creatures, but sometimes the insects' light organs were removed and transformed into streaks of glowing war paint on our faces. More often than not the captives in the mason jars stopped flashing after a while and turned belly-up. It sounds awful now. If I had known then what the fireflies were doing on those summer nights, I would have opted out of the backyard carnage.

Fireflies generate their light by creating chemical reactions between luciferin, luciferase, oxygen, and ATP, which is made by the mitochondria of the firefly's cells. Fireflies do not use bacteria to produce their light, but they have incorporated into their DNA the same genes used to make luciferin and luciferase. The firefly's light organ is very efficient. Nearly 100 percent of the energy produced in the organ is transformed into light. In contrast, only 10 percent of the energy flowing into an incandescent lightbulb is used to make light; the remaining 90 percent is wasted as heat.

The flying, blinking fireflies are all males. Females are wingless, so they sit on top of the grass waiting for the right signal to come along. The language of fireflies is coded in the patterns of flashes seen on a summer evening. Each species has a unique code that only its members understand, with the flash patterns of the males and females being distinctly different. *Photinus consimilis,* native to the Ozarks, cruises across the lawn while flashing coded signals in hopes of attracting the attention of a female and receiving a coded response. For firefly males, as for most species, attracting a mate is never easy. Some male fireflies flash for up to three weeks before getting a reply. The signal of the male Ozark species is given as a series of relatively rapid flashes. When a female sees a signal that she likes, she delays her reply for a specific length of time that has been programmed in her genes and then flashes once. The female's flash is a beacon that reveals her location to the male of her choice. If she is really impressed with a male's performance, she emits up to a dozen pulses to make sure that male finds her.

Marc A. Branham, a young scientist from the University of Kansas, determined the species-specific flash pattern of *P. consimilis* by videotaping its flash patterns, analyzing the video, and developing a computer program that could reproduce the pattern with a tiny light diode. He took his computer-programmed diode back into the wilds and set about seducing *P. consimilis* females in an area not far from where I grew up, known as Roaring River State Park. After trying out various rates of flashes similar to those he had observed on the videotapes, Branham locked on to the *consimilis* code and discovered that the females were most attracted to males that could flash faster than their competitors.

The male must build his levels of luciferin and luciferase to a critical mass rather quickly and energetically to be able to display a series of rapid pulses, which are similar to the flashes of the older battery-powered units mounted on traditional 35 mm cameras that had to recharge for a few seconds between flashes. Males that can flash faster are likely to be fitter than fireflies of the same species that pulse at slower rates. After the female makes her choice, and assuming the male has been able to reach her in time, they will mate and the female will lay her eggs on the soil. After about a month, the eggs hatch and produce larvae that are voracious predators. The larvae, commonly known as glowworms, already possess a light organ that flashes when they are disturbed by a predator. The message is "I contain chemicals that taste bitter. Do not eat me." The larvae feed on insects, insect larvae, including other firefly larvae, and snails until just before the first frost in the fall. The frost itself is a signal that tells the larvae to burrow into the ground and remain there until the spring. The larvae emerge and feed until early June, when they pupate for two and a half weeks and transform into adult fireflies.

But firefly mating is fraught with eavesdropping predators and cheats. As the firefly males are competing with their signals, they are keeping a keen eye out for the come-hither flash of a female. When a female makes her response, all the nearest males rush to her location and attempt to mate with her even though her signal was intended for only one. Scientists call this a scramble.

Thomas Eisner, of Cornell University, one of the world's leading experts on insect communication, has discovered a deliberate, rather ghastly form of deception taking place on those warm summer nights. Female fireflies of the much larger genus *Photurus* have learned to imitate the mating signals of *Photinus* females. These femme fatales are lurking in the yard, waiting to lure unsuspecting males over for dinner. While *Photinus* males are flying over the yard, signaling and searching for the welcome flash of a *Photinus* female, the *Photurus* females mimic the *Photinus* female's code. The naive male that accepts this deceptive invitation is met by the open jaws of the *Photurus* female, which proceeds to eat him alive.

According to Eisner's research, *Photurus* is after more than a meal. *Photinus* males possess a valuable substance in their blood, known as lucibufagans, which repels predatory spiders, birds, and bats. Whenever the *Photinus* males are disturbed, they excrete a little of this protective fluid, which predators of other species find unpleasant. Eisner, David E. Hill, Scott R. Smedley, and Jerrold Meinwald collaborated on a study of this "murder and mimicry" and discovered that the protective chemical in the blood of *Photinus* males is similar in composition to the bufalin steroids contained in the venom of poisonous Chinese toads. The scientists came up with the name lucibufagans, derived from the Latin words for *light* and *toad*. (Because I am slightly dyslexic, I thought until very recently that the word was *lucibugafans,* which I had thought was an intentionally humorous name, unusual for scientists.) The *Photurus* female eats *Photinus* males to obtain their lucibufagans. *Phidippus* jumping spiders are a predator of the wily *Photurus* female. They will eat those that have not dined on a *Photinus* male and avoid those that have ingested the *Photinus* lucibufagans. When the scientists painted lucibufagans on fruit flies, which do not normally possess it, they too were protected from predators. "This strategy of acquiring ready-made defensive chemicals from other organisms turns out to be quite common in nature," Eisner says.

Eisner discovered that the male fire-colored beetle secretes a chemical from a gland in its forehead that protects it from predators. When the male beetle goes searching for a mate, it will secrete a little sample of the protective fluid and offer it to a female as a nuptial gift. Females that accept the gift and consent to mating receive the chemoprotection for themselves. The male's sperm, which also contains the chemical, confers protection on the female's valuable eggs. Just as the *Photurus* female gains chemoprotection by eating male fireflies of a different species, the male fire-colored beetle obtains the superjuice for its nuptial gift from eating dead blister beetles or blister beetle eggs. Another common name for the blister beetle is Spanish fly. The fire-colored beetle is seeking a compound in the blister beetle called cantharidin, which happens to act as a potent vasodilator in humans, similar to Viagra. The cantharidin of the Spanish fly became a legendary aphrodisiac in the 1800s after a rather unfortunate incident that occurred among some French Foreign Legion troops. The troops had been dining on frogs' legs, and the frogs, unbeknownst to them, had dined in turn on blister beetles. Within a couple of hours after dinner, the troops sought out the regiment physician, complaining of prolonged and painful erections.

The most important signal for many species of insects and mammals is the pheromone, whose power has become almost mythical. The word *pheromone,* coined in 1959, is derived from the Greek words *pherein* (to transfer) and *hormon* (to excite). Pheromones, which are sensed through receptors in the olfactory organs of animals and insects, can seduce and even command members of the opposite sex to mate. In human couples, they may help insure genetic compatibility and may even affect fertility. Pheromones are at work in all species, but none make broader use of chemical signals for communicating than the insects, in which humans first discovered them.

A pheromone is any substance secreted by an organism outside its body that causes a specific reaction in another organism of the same species; pheromones are divided into two types: releaser pheromones, which command an immediate behavioral response, and primer pheromones, which induce a longer-lasting physiological response in the receiver. The first pheromone discovered was in a moth. After purifying a small amount, scientists could induce a male moth to beat its wings in a sexual frenzy by wafting the tiniest fraction of it his way. Females release a pheromone that can be detected by males up to about three miles away. Males have developed over-sized antennae specifically for picking up the signal of the female moth from such great distances. Feathery antennae increase the surface area for signal detection. The *Danio* butterfly has evolved a strategy that makes it look a bit like a fighter jet being refueled in midair. The male locates a female, hovers above her, and extends a brushlike structure toward the female's antennae to make contact. The brush is covered with a male pheromone that her antennae are tuned to recognize. If the male's signal excites, she responds by landing so the two can mate.

Entomologists generally divide pheromones into nine functional varieties: sex, aggregation, dispersal, alarm, recruitment or trail following, territorial or home range, service or job description, funeral, and invitation. Queen bees, ants, and termites possess an excessively large amount of pheromones used to attract drones for mating and to identify colony members. Queens can release pheromones that compel males to mate with them, prevent subordinate females from ever mating, and keep colony members working around the clock. Queens are totalitarian rulers.

Both animals and insects have evolved special glands to secrete chemical signals. Insects have an astonishing variety of abdominal glands dedicated to marking trails—termites have at least nine. Entomologists have identified 85 different glands that secrete chemical messages: 39 in Formicidae, 21 in Apidae, 14 in Vespidae, and 11 in Isoptera. The various functions of these glands correspond with their location on the insect's body. Alarm pheromones are secreted from mandibular glands in a variety of ants and bees, and from a frontal gland on the anterior portion of the head in termites. An alarm pheromone is released to rally troops. When a wild chimpanzee places a twig in the entrance of an ant colony, the sentries near the entrance release an alarm pheromone that causes the troops to come running. The chimpanzee patiently waits until the ants are swarming the stick and then eats them as if they were a Popsicle.

Termites and ants frequently wage war against each other. Termite soldiers carry a potent ant repellent in their anterior glands and release it as a chemical weapon on the battlefield. To the ants, it is a weapon of mass destruction.

The large number of secretory glands found in social insects suggests that the type of information being conveyed via chemical signals is rather complex and may provide a fair amount of detail about the location and quality of a food source for which a trail is being marked, for example. This is most certainly the case in ants and in honeybees, which have also developed an elaborate communication dance.

One ant species, *Dinoponera quadriceps,* does not have a queen, per se, but it does have an alpha female. All the females in the colony are workers that are capable of breeding, but the job of reproduction is monopolized by the alpha female, called the "gamergate." *Dinoponera* colonies are small compared with those of other ant species, with only about 80 adult female workers. A linear hierarchy exists, with three to five high-ranking workers, called betas, which do little work. These betas are runners-up for the alpha position should old age or some "accident" result in the death of the gamergate. The males have no role in the social structure of the colony, except that a single male will be selected by the gamergate for mating. The workers, as well as the ladies-in-waiting, typically are daughters of the gamergate. Sometimes one of the betas seems to grow impatient and challenges the gamergate to a fight during which they chase each other, wrestle, and wreak havoc throughout the nest, trampling on anything that gets in their way. During these bouts, the beta and the alpha will take short breaks, at which time the alpha will attempt to rub the beta with her stinger, which disperses a powerful chemical signal that induces the low-ranking sisters of the beta to grab and immobilize her. Sometimes, perhaps in the heat of the moment, the beta is killed. If the beta survives the immobilization, she is demoted from beta status and one of the low-ranking sisters will replace her. The scientists say the behavior "precisely fits the definition of punishment." Had the pretender to the alpha position waited, much as Prince Charles has been waiting for the crown, she might have ascended one day and spread her own genes through the colony. But the punishment for the failed coup is never to be in a reproductive position again.

In mammals, pheromones are excreted from endocrine and apocrine glands. An unspayed housecat releases a pungent pheromone and rubs it on the furniture to advertise her readiness to mate. Male boars and hedgehogs secrete their courtship pheromones through frothy saliva. Guinea pigs, rabbits, agoutis, and chinchillas mark female mates by urinating on them, and females return the favor.

Chemical signals can be dispersed into the wind or a current of water by simple diffusion, often assisted by filaments, such as those in moths, that extend over a secretory gland. In humans, the hair of the armpits and pubic area are believed to disperse sex pheromones, with, of course, the assistance of bacteria. Some animals prefer dispersing their pheromones by rubbing secretions onto the body like perfume.

Jack Bradbury and Sandra Vehrencamp, animal communication professors at Cornell University in Ithaca, New York, have uncovered dozens of unusual strategies that animals have adapted for dispersing pheromones. (Bradbury and Vehrencamp's textbook is the most widely used in animal communication courses in the United States.) The male white-lined bat positions itself in front of a female during flight and hovers while wafting his pheromone at her. Researchers had thought that the bat's wings contained scent glands, but Christian Voigt recently discovered that the male bats have sacs in the wings in which they deposit various secretions to manufacture their cologne. Every day at about 3 p.m., the males begin grooming themselves in preparation for the big seduction. The bat licks his penis and gets a small amount of a secretion on his tongue and places it in the wing sac. Then he adds some urine. The mixture reacts with bacteria that colonize the sac and, voilà, bat cologne.

Both male and female Douglas fir beetles employ pheromones in their nuptials in an elegant but complex manner, according to James Gould of Princeton University. Attracted to the tree by the smell of its sap, the female lands and excavates a small tunnel. Trees that have been injured but are still living are more likely to exude sap and attract the female. Ironically, the tree's sap is its best defense against other burrowing insects. The female beetle carries with her a blue-stain fungus that she deposits in the burrow. The fungus grows and plugs the pores from which the tree's protective sap would normally be oozing to ensnare the beetle. The female releases a pheromone signal that summons other females as well as males to the tree, because the attraction strategy works best when many females burrow at the same time. The pheromone signal is ten times more potent in its odor than the tree's sap. Once all the females have arrived, they simultaneously burrow and make a clicking sound that resonates inside the tree. This audio signal coordinates the spacing of the burrows and prevents the beetles from burrowing into each other's nests. When males arrive at the burrows, they too create vocal signals, which identify their species, and then pair off with the females. The males then release a pheromone that signals to late arrivers that there is no vacancy in that burrow. If a male ignores the keep-out signal, the resident male issues a low-frequency warning sound. If pushed, the resident will battle the intruder. In the meantime, similar "no vacancy" pheromones are being released by all the new male residents, which create a dispersion cloud around the tree, a group signal that deters any other arrivals. This also prevents too many beetles from burrowing and killing the tree, since the larvae that will hatch from the eggs in the burrows need live tissue on which to feed.

The following mate selections are a notable oddity in the insect world. Even though females usually prefer dominant males, Allen Moore of the University of Manchester, in Britain, discovered that female Tanzanian cockroaches assess male sex pheromones and choose partners that have smells associated with the lowest social ranking. The reasons for this odd choice of partners are unclear. Females that mate with low-ranking cockroaches produce fewer male offspring than if they mated with the top bugs, which goes against the notion of trying to get as many genes as possible into the gene pool. Moore speculates that by producing fewer males, the females reduce competition among the male offspring and thereby increase the odds that the sons will find mates.

Chemicals help many species recognize their own family members. Recognizing offspring and other kin by olfactory signals is important for maintaining order among social creatures. Mammal mothers can usually identify their newborn offspring by scent alone. Knowing who one's nest mates are is critical for defending colonies from intruders. When ants release the alarm pheromone and catch an intruder, they tear it apart, limb by limb.

The natural pheromones of humans appear to be produced primarily by apocrine sebaceous glands in the skin that are concentrated in the underarms, the nipples of both men and women, and the pubic, genital, and circumanal regions (the outer pigmented area of the anus) and around the lips, the eyelids, and the outer ear. Humans beings are believed to secrete pheromones in response to fear, sexual arousal, stress, and many emotional states.

Males in mammal species wear the scent of either social dominance or submission. The stronger or more powerful the scent, the more dominant the male is likely to be. Before sexual maturity, male elephants exude a liquid from beneath their eyes that smells a bit like honey and actually contains compounds found in honey (it attracts bees to the young elephants like mad). As they mature, the elephants' smell changes to a more powerful skunklike odor, and then, as one scientist puts it, to "the odor of a thousand goats." The older and more dominant a bull becomes, the worse he smells to humans, but female elephants appear to find the odor quite attractive.

As a rule, females prefer the smell of dominant males, but male mammals generally appear to be attracted to the scent of any female. Rodents are considered to be particularly vulnerable to the effects of pheromones, and in mice, the mere whiff of a female sex pheromone can compel a male to mate at once.

Alan Singer, of the Monell Chemical Senses Center in Philadelphia, and Foteos Macrides, of the Worcester Foundation for Experimental Biology in Worcester, Massachusetts, conducted a study in golden hamsters to investigate how strongly rodents might be influenced by sex pheromones. They anesthetized a male and put it in a cage with a male hamster that was awake. The alert hamster either chewed on the other guy's ear and dragged it around the cage or ignored it. This served as a baseline of normal behavior for a caged alert male hamster in the company of an anesthetized male hamster. In the next part of the experiment, the researchers spread the vaginal secretions of a female hamster on the anesthetized male and then sat back to see what happened. The alert male became highly aroused by the female scent and attempted to mate with the anesthetized male.

Mammals detect pheromones through the vomeronasal organ, or VNO. It is located behind the nostrils where the nasal septum joins the hard palate, but it has nothing to do with the normal sense of smell. In mammals, the VNO looks like two little pits in humans, not unlike the pits in the fer-de-lance snake, and while humans appear to possess only vestiges of the organ, a growing body of research indicates that it is still functioning. In mammals, the VNO commands its own exclusive nerve pathways to the brain, bypassing the main olfactory bulb and plugging into something called the accessory olfactory bulb. Sex is so important to nature that it provided specialized neural wiring for high-speed access to pheromone signals. From the accessory olfactory bulb, nerves lead directly to areas of the brain associated with reproductive and maternal behavior. The VNO is regarded as an extremely ancient, highly specialized organ.

All primates appear to rely, to some extent, on olfactory signals to attract mates and to determine social hierarchies. In 1971, Martha McClintock, then at Harvard University, conducted a classic study on the "dormitory effect." She showed that women living in close quarters—in this case a college dormitory—tend to have their menstrual cycles at about the same time. This finding was quite surprising. The synchronizing effect is attributed to a pheromone in the women's underarm odor. In the wild, some dominant female mammals exude a pheromone that synchronizes the estrus cycles of subordinate females with their own. This helps ensure that other females do not usurp the dominant female's position by gaining favor with dominant males during periods when the dominant female is not receptive. McClintock's study suggests that, just as animals do, dominant women, via pheromones in their sweat, are able to synchronize other women's menstrual cycles with their own. We can call this the queen bee effect.

Michael Russell decided to examine the queen bee effect after one of his female friends mentioned that the menstrual cycles of her friends tended to synchronize around her own. This gave Russell an idea for a simple but clever study. He persuaded his friend to wear sterile pads under her arms for a while and then recruited 16 unsuspecting women for his study. The women came to his lab three times a week for four weeks. Russell had made an extract of his friend's underarm sweat from the used pads and mixed it with alcohol; he applied this to the upper lips of half the women during their visits. He applied an alcohol-only solution to the upper lips of the other women. By the end of the study, four of the women receiving the underarm sweat applications had synchronized perfectly with his friend's cycle, even though they had never met her. The other four women in this group also fell closer in line to the dominant female's cycles. The women in the alcohol-only group showed no effect.

After the study was concluded, Russell went on to look at whether humans can recognize their own scent and whether they can tell the difference between the scents of males and females. He recruited 16 male and 13 female students and had them wear clean T-shirts for 24 hours. No one was allowed to use soap, deodorant, or perfume. After 24 hours, the T-shirts were removed and the students were asked to smell the armpits of the shirts. Results: 81 percent of the males and 69 percent of the females were able to identify their own odor. Similarly, most males and females were able to tell the difference between the T-shirts of males and females. The students described the smell of men as "musky" and the smell of women as "sweet."

A team of scientists at the University of Texas at Austin asked 17 women to sleep in the same shirt three nights in a row during their peak period of ovulation and to wear a different shirt during the least fertile phases. They could not use soap, perfume, have sex, or take birth control pills during the study. When 52 men were asked to sniff the garments and rate how sexy they smelled, they best liked the way women smelled during peak ovulation.

Studies suggest that women are sensitive to the musky smells of men and that their sensitivity increases as they near ovulation. What is it in a man's sweat that has such an effect on women? George Dodd, of the University of Warwick in British Columbia, Canada, says the ingredient is the male steroid androstenol, which, he says, has all the qualities of a sex pheromone. Dodd has received a goodly amount of media attention, including coverage by the *BBC News* and *Time* magazine. Dodd conducted a study of 76 male and female volunteers in which women were given brief whiffs of androstenol, which he said made the women more likely to be social with men afterwards. The whiffs had no effect on the men.

Dodd claimed that androstenol, which he called Osmone 1, may be a precursor to the male hormone androsterone. It allegedly has soothing effects and is sold commercially as aromatherapy. Dodd has worked as a consultant to a British biotechnology company that is manufacturing commercial pheromone-based products. According to Dodd, Osmone 1 falls into the general category of musk. Women are reportedly 1,000 times more sensitive than men to the odor of musk. That sensitivity, however, appears to be blunted in women who take oral contraception and in post-menopausal women.

Many scientists remain skeptical of commercial products that claim to have captured the essence of sex pheromones and promise to lure members of the opposite sex. If these products were valid, one might have to consider the ethics of such products in human society. In the animal kingdom, according to Bradbury and Vehrencamp, "Pheromones have a greater potential for manipulating a potential mate or competitor to the benefit of the sender and to the detriment of the receiver because they can modify the receiver's internal hormonal state."

The musky smell of men has been most strongly associated with testosterone and its precursor chemical, known as androsterone. In the animal world, the equivalents of these steroid hormones act as both sex pheromones and indicators of social status. Certain fatty compounds in vaginal secretions and chemicals in the urine of female animals are indicators of sexual receptivity and potent stimulators of arousal in males. In an earlier chapter, I mentioned that a bull elephant samples a female's urine and places the tip of his trunk against his VNO. If she is in estrus, her pheromone signals his brain to have an immediate erection. After mating, the female delivers another powerful chemical to the male, which blocks the effects of nitric oxide and causes him to lose the erection.

Sex pheromones are the heavy artillery in the battle of the sexes. They can instantly arouse members of the opposite sex and, most important, they help females make the right choices of mates. Nature has invested much in the development of pheromones and the VNO. Studies show that the animal's immune system type is among the different kinds of information contained in a pheromone. Of particular interest is a set of genes called the major histocompatibility complex, or MHC for short. The MHC genes first came to the attention of researchers who were trying to solve problems of rejection among recipients of organ transplants. MHC types are analogous to blood types, and matching an organ donor's MHC type with the recipient's can reduce the chances of rejection. MHC genes are involved in immune system responses to infections, helping to marshal infection-fighting T cells. In animals, MHC genes are linked to the ability to resist parasitic infections. Female house mice are able to determine the difference between healthy males and males infected with parasites based on their olfactory senses. The female's VNO allows her to select the healthiest males for mating, increasing the odds that her offspring will be able to resist parasites.

MHC genes may also play a role in ensuring genetic diversity in species, allowing animals to assess MHC types through odor. Mating pairs with similar MHC genes produce offspring with less genetic diversity than pairs with dissimilar MHC genes. Inbreeding and loss of genetic diversity can result when close relatives with quite similar MHC genes mate. Studies confirm that odor types are directly linked to MHC genes. According to Kunio Yamazaki and colleagues at the Monell Chemical Senses Center, the MHC genes themselves code for the pheromones that allow animals to choose mates with dissimilar types. This suggests that potential mates can sniff-test for mates that will provide their offspring with the greatest genetic diversity. Such pheromones would be strongly favored by natural selection and perhaps sexual selection as well.

Female stickleback fish prefer males whose scent is associated with MHC genes that are the most different from their own, according to Thorsten Reusch, of the Max Planck Institute for Limnology in Plon, Germany. Yamazaki found similar results in mice and suspects the same to be true of humans. In an interview with the journal *Nature,* Reusch said, "It is very likely that many other organisms have the same strategy." MHC studies in humans are less clear in their conclusions, but they are suggestive. Among the most recent research is yet another smelly T-shirt study by Carole Ober and Martha McClintock, who pioneered the technique, at the University of Chicago. This time women were asked to smell two-day-old T-shirts worn by men with different sets of MHC genes. The shirts were hidden in a box that contained faint household odors, such as bleach and fresh laundry. Women were asked which smell they could tolerate best for the rest of their lives. In this case, the women chose the smell of T-shirts from men who had MHC genes similar to their own. Based on the previous studies, one would have predicted that the women would have selected T-shirts representing genes that were more different from their own. According to McClintock, females inherit their MHC genes only from their fathers. She attributes the discrepancy to the possibility that women are selecting partners that represent a more intermediate blend of MHC genes. Some human fertility experts are beginning to suspect that incompatible MHC genotypes may be a cause of frequent miscarriages or the inability to conceive in a couple who seem to be potentially fertile. The evidence for this is clear in animals, including fish.

The importance of MHC typing is not lost on the perfume industry. Manfred Milinski and Claus Wedekind, of Bern University in Switzerland, studied people's taste in perfume and its relationship to MHC gene types. They concluded that a woman's preference for various perfumes and aftershaves is associated with her MHC type. The researchers suggest that a person's taste in perfume might serve as a simple proxy for MHC gene types. That seems to be a stretch. But if a person discovers that the sweat of a member of the opposite sex is attractive in a way he or she cannot describe, perhaps it is the chemistry of love, and a sign of genuine (MHC) compatibility.

The game of love relies heavily on the olfactory sense, but with the exception of the amazing moth, getting physically close enough to employ one's chemical weapons against a member of the opposite sex is by no means a simple matter for most species. Males and females have different goals, which would suggest that the so-called battle of the sexes is indeed a real battle. It begins at the level of the sperm and the egg.

According to Bradbury and Vehrencamp, "Males are less choosy and more eager to mate than females because their investment in gametes is lower, so they play the active, pushy role during courtship. Errors in species recognition and mate choice are more costly for females so they evaluate the male carefully and signal their acceptance (or rejection) of the mating."

The evolution of mate attraction and courting signals is tied directly to the economics of sperm and egg production. Sperm are a dime a dozen in males, which leaves males with energy to invest in expensive visual and vocal signals to attract a female. Sperm is also easy and cheap to resupply for a fast turnaround, so males can afford to mate more often. Male lions can copulate more than 100 times in a 24-hour period.

Eggs, however, are limited in number and available for fertilization only periodically. Females spend a significant amount of their energy on their hormonal cycles, which leaves less energy for developing colorful, elaborate visual and vocal signals. Along with these energy-demanding hormonal cycles, the females of many species, including monkeys, horses, elephants, moths, butterflies, lizards, snakes, rodents, and deer, have evolved inexpensive and powerful olfactory signals. The female's chemical repertoire is manufactured with great efficiency from the hormones that are circulating at high levels during ovulation, and it is generally more potent than that of the male.

The dramatic difference in the costs of egg and sperm production sets the stage for truth in advertising during courtship in the animal kingdom. In human society, courtship is a period of great deception. Men and women tend to be on their best behavior and may be inclined at times to say whatever they think the other wants to hear, even if it means exaggerating or lying. The conversations that take place between courting animals have evolved to be much more honest, under direct pressure from the females of most species. If the receivers, which are usually females, were not so discerning about the large number of advertisements they receive from males, they would squander their very expensive eggs and the energy invested in their hormonal cycles on males that may be unhealthy or lack the dominance and maturity to secure territory and food or provide parental care when required. Some of my friends complain about responding to personal ads only to discover that the other person has engaged in false advertising about their traits. The honesty demanded by females in the animal kingdom enhances their odds of mating with the most desirable males and passing along to their offspring the best chances of survival.

Pheromones that convey information about the MHC complex and other physiological traits are inherently honest, but beyond chemistry, females are also looking for truthful advertising about the genes a potential mate carries to ensure they are healthy for her offspring. How that health is reflected depends on the species. It may be the quality of a songbird's voice and the size of his song repertoire, or the brightness of the colors in his feathers. In other species, health may be conveyed by the deepness of his voice, his material wealth or his size and dominance.

Studies have shown that the bright colors of the feathers of male birds, while appealing to the eye, are dependent on a diet rich in carotenoids and so indicate good health. Dull colors on what should be a brightly feathered bird reveal illness or lack of success in competing for food with other males. The ability to sing a song longer than another male, or sing more songs, or display longer and more vigorously signals greater endurance. The more it costs a male to produce a signal, the more likely it is to be an honest one, according to Bradbury and Vehrencamp.

The high cost of signaling becomes readily apparent among males that must vocalize loudly to grab a female's attention. These males run the risk of being overheard by a predator and tracked down for dinner. The classic example used by many scientists to illustrate the relationship between cost and honesty is the Tunagra frog of Central America. Its typical territorial call is an elongated whining croak, which makes the frog difficult to locate by predators, such as the fringe-lipped bat, because of the way the sound propagates in the surrounding environment. A long call that spreads out, the whine of the healthiest male Tunagra frog does not impress the female, which insists on a more complex and costly low-frequency chucking sound for courtship. When males are competing among themselves, they vocalize only with the whine, but when females arrive to assess the males for mates, the males change their song from "whine, whine, whine" to "whine, chuck, chuck, whine." The low-frequency chuck honestly reflects a male's body size and overall health. Females choose to lay their eggs next to the male that makes the loudest and deepest chuck, but the eavesdropping fringe-lipped bat also has no difficulty locating the chuck because the shorter burst of sound produces an "I'm over here!" quality. Risking its life, the male will not produce a deceptive signal that may be ignored by the female.

Honest signals, or truth in advertising, also account for the exaggerated physical characteristics, or ornaments, that males have evolved to dazzle potential mates—a phenomenon known as the handicap principle. The peacock's long, colorful train of feathers is a perfect example of ornamentation that signals honesty. The peacock with the longest, most colorful tail feathers is the fittest by virtue of his ability to support the otherwise cumbersome, useless ornament, which can even be a handicap should a fox decide to chase him down for dinner. Amotz Zahavi proposed this handicap principle in 1975, and it is widely accepted as a rule by animal communication scientists. In Zahavi's words:

> An individual with a well developed sexually selected character (such as a peacock's flashy tail) is an individual which has survived a test. A female which could discriminate between a male possessing a sexually selected character, from one without it, can discriminate between a male which has passed a test and one which has not been tested. Females which selected males with the most developed characters can be sure that they have selected from among the best genotypes of the male population.

The handicap principle assumes that an unhealthy male cannot afford to invest his limited resources into making such an elaborate ornament. The notion behind this is borrowed from Thorstein Veblen's 1899 book *The Theory of the Leisure Class,* in which he coined the term *conspicuous consumption.* Veblen was particularly interested in the implications of women's flamboyant Victorian costumes. According to Veblen, only the truly rich could afford to flaunt their wealth. Animal communication scientists have applied this concept to the evolution of signals to attract mates. Only the fittest can afford to flaunt elaborate ornaments and other signs.

In the next two chapters, we'll look more at the songs and visual displays of mammals and birds, which, although they may vary widely from species to species, have much in common. Some general rules of the natural language for attracting a mate in the wild world of animals are:

1. Give the females what they want, which is males with good genes and, depending on the species, a nice spot of well-defended territory, plenty of food, and a little help feeding mother or the babies.

2. Give the males what they want, which is females. Usually, any female of the same species will do, and the more the merrier.

3. Females do not need to try hard. There are a few species whose females advertise for males, but they basically stay in one place, flash a light organ (fireflies), vocalize with a long-distance call (elephants), or release pheromones by rubbing them on bushes, urinating on the ground, or sending them adrift on a breeze or water current. Males will come running, flying, or swimming from all directions. Sometimes it boils down to a scramble: the first male to arrive wins. More often the males compete directly, and the dominant male wins.

4. When males advertise, they often stay in one place and make loud calls that convey strength, good health, and territorial dominance. All long-distance calls use low-frequency sounds, which travel farthest in air and water. Or, in species that rely on visual signals that must be seen at a distance, they may use different types of bright colors for creating contrast against the natural background.

5. Females will signal acceptance clearly. Faced with a choice, the female signals to her preferred mate that he is acceptable in body language that is unambiguous, using intention movements associated with the normal movements

made during copulation, for example, or by releasing pheromones to indicate sexual receptivity.

6. When a female approaches, males must respond appropriately. The male should display strongly but not too strongly. Vocal males should reduce the loudness of the call. As the female draws nearer, males must increase the rate of the signal, whether vocal or visual, to indicate genuine interest in the female. Females are often skittish at the beginning of courtship and may be fickle, so the male has to read the female's body language carefully to know which way to go. At close range, he must draw attention to any assets that will impress the female, such as a nest. Finally, he must release pheromones. Depending on the species, some males should urinate on the female to aid her olfactory assessment of his qualities.

With these general rules in hand, let us move on to the many love songs that are to be heard in nature. Music was invented, after all, by the animals.

Six

SONGS AND SHOUTS

A NDREW BASS and his wife, Margaret Marchaterre, of Cornell University are on a most unusual fishing trip. They have pitched a tent on the rocky shores of Washington State and cast a line into the cold Pacific Ocean. Instead of fishing line, Bass is using a heavy-duty wire similar to the type used to connect stereo speakers to a receiver. And instead of bait at the end of the line, Bass is using a hydrophone.

As night falls, blanketing the sky with thousands of blinking stars, Bass lights a Coleman lantern and adjusts the gear. The scientists are fishing for the sounds of a rather homely aquatic species called the midshipman, which earned its name for the row of bacteria-assisted bioluminescent disks that run along its side, like the buttons on an old-fashioned naval midshipman's dress coat. The species is an ideal candidate for studying communication among fish. It sings. To a female midshipman's ears, the loud low hum of the love song must sound like sweet music. The sound reminds Bass of the chanting of Tibetan Buddhist monks. To others, the songs are more reminiscent of a foghorn or the drone of an old propeller airplane.

The males of the midshipman, as with many species of fish, are the nest builders. They prefer the safety of tidal pools near the shore where they excavate shallow cavelike nests with their tail fins. From the nests, they hum to attract females to come and lay their eggs. The low-frequency sounds travel through the water and can attract female midshipman from more than a mile away. The underwater serenade begins after nightfall during high tide. Bass has recorded as many as 30 midshipmen singing in unison. The combined sound can be as loud as a fleet of motorboats. If the song could be translated into words, it might say, "Come to me. I am strong. Our offspring will be great survivors." The females gather at a short distance from the choir of males, from where they choose the singer they find most attractive. Bass and one of his graduate students, Jessica McKibben, have discovered that to focus on individual songs, the females employ a trick that humans call the cocktail party effect.

Only the best singers can lure the female midshipmen to their nests. Once the female chooses a male, she will spend up to 20 hours laying about 200 eggs on the roof of the nest. The male fertilizes the eggs with his sperm one at a time. When he has finished he will resume singing to keep the females coming. A successful crooner will attract about 15 females to his nest and fertilize about 3,000 eggs.

The female's ability to tune in selectively to individual singers while masking the sounds of the others has implications for neuroscience and human brain function. The results of studying these fish will allow scientists to understand human brains better and perhaps isolate similar areas of the brain used by people to focus on information, stimuli, and conversations they want to hear or take in while ignoring everything else.

Before traveling to the Washington shore, Bass and McKibben created computer-synthesized sounds in the laboratory that mimic the songs of various males, to learn the qualities female midshipman are looking for in a song. They played the synthesized songs over multiple loudspeakers in an underwater tank to determine which songs a female midshipman preferred, and attracted the female directly to the loudspeaker that was playing her favorite song. The female would nuzzle the speaker and swim circles around it, searching for the singer and a nest. Finding neither, she eventually swam away without releasing her eggs. From these playback studies, Bass learned that female midshipmen choose their mates based on the song's pitch, duration, and loudness. They like them sung low, loud, and long. As with the bullfrog, the pitch of the song correlates with the body size of the singer. Most of the males' love songs last only a few minutes, typical of human songs. But Bass recorded one supercrooner in the wild that hummed for more than two hours.

The vocal signals animals use for mate attraction have evolved so that they are broadcast on frequencies unique to the males and females of a given species. The songs of midshipman males are broadcast only on HUM 100 on the FM fish dial, and female midshipmen are tuned only to that station. The sounds of other species, unless they happen to be a predator, are usually background noise to a midshipman's ears. Bass's studies of the brains of male midshipmen reveal that they have a set of motor neurons in their hindbrains that produce signals at frequencies of 90 to 100 Hz. These neurons send a signal to a set of muscles in the gut that vibrate a bladder normally used for buoyancy. A second set of neurons in the brain stem translates the signals into the instructions that tell the motor neurons and the muscles what to do. Reflecting nature's economy, the same area of the human brain stem controls the larynx, which is used for speech. Female midshipmen have evolved a set of neurons in the midbrain that detect songs sung at frequencies of 90 to 100 Hz.

When a male midshipman that is singing discovers another male has invaded his territory to fertilize the eggs in his nest, the singer will stop humming and growl at the intruder. In territorial conflicts, fish make low-frequency grunts in a staccato pattern similar to that of mammals and birds. The invaders, called sneaker males, are smaller than the singers and do not sing or build nests. They have evolved to resemble females, a disguise that allows them to approach a singer's nest. Then they suddenly dart in to fertilize a few eggs of their own. These small populations of cheaters survive because they do not upset the status quo of the midshipman species.

Sneaker males can also be found in gobies, sunfish, squid, bullfrogs, and even elephant seals. This alternative mating strategy evolved out of the intense competitive pressure among males. Scientists believe that the majority of males among animals never find a mate because females tend to select the dominant males. Scientists estimate that only about 1 percent of elephant seal males, for instance, are able to attract a female and mate. Dominant male elephant seals acquire harems that they vigorously defend. Smaller male elephant seals, which have little chance of winning a battle with a dominant male, hang out on the periphery of his territory and wait for an opportunity to slip in and mate. Among terrestrial species, from songbirds to deer, young males typically resemble females until adulthood. The female appearance may occasionally help a sneaky youngster gain access to a female, but the phenomenon most likely evolved to lower aggression between older, stronger males and younger, weaker males. Otherwise, many of the adolescents might not survive to become adults.

The grand musicians of the sea are the humpback whales. Ancient mariners were most likely describing the songs of humpback whales when they talked about the mythical sirens of the sea. In the 1960s, marine biologist Roger Payne came up with the idea of dropping an underwater microphone over the stern of his sailboat to try to capture the sounds of these sirens on tape. Most people at the time thought the idea was ridiculous, but the experiment was an incredible success. In 1970, Payne produced a record album, *Songs of the Humpback Whale,* that became a hit. The album drew the public's attention to commercial whaling, which was threatening to wipe out many species. The following year, Payne and Scott McVay published the first report on the songs of humpbacks, which became a hit with the public, attracting hundreds of men and women to the field of marine mammal communication.

Since the first experiments, Payne has reflected, "Some of my happiest hours have been spent at night lying back in the cockpit of a sailboat, alone on watch, steering with one foot and watching the mast sweeping across vast fields of stars, while the songs of the humpback whales poured up out of the sea, to fill my head, my heart and finally my soul as well."

According to Payne and Peter Tyack, a leading marine mammal communication scientist at the Woods Hole Oceanographic Institute on Cape Cod in Massachusetts, humpbacks produce the longest, most varied songs in the animal world. The haunting vocalizations are described as song because they follow a series of different themes in a predictable order. Humpback whales invent a new song each year that contains some parts of the song from the previous year. As time passes, and new elements are added, the original song fades away until it is no longer recognizable. Individual whales sing their own version of the common theme.

Studies of the humpback's songs reveal that they contain repeated phrases that combine to form the themes within a song. Themes are combined in patterns that become the song. On average, a humpback song will contain three to nine themes and last from 8 to 15 minutes. Some songs may last as long as half an hour. Whales repeat their songs over a period of several hours. They must surface to breathe, so they take quick breaks about every 15 minutes. Singers that can last the longest between breaths may have a competitive advantage over singers that do not have as much stamina.

As the common song is always changing, a new one usually emerges just before the mating season and becomes the number one hit for the season. One hypothesis is that the older dominant males may influence the style of the new song, which may be imitated by the younger males. After all the males in a region have begun singing the same song, the older males will begin inventing new variations to stay a step ahead of the competition. Tyack compares the competition among humpback singers to the rivalry among pop music composers vying for the *Billboard* chart's Top 10.

Since the whale songs change from year to year and the hit single evolves during a mating season, scientists say it is doubtful that songs contain any semantic information comparable to lyrics. According to Tyack, the main purpose of the song seems to be to manipulate the behavior of females and other males. To manipulate the behavior of another whale, the song needs only to convey the qualities of the singer. For example, it is not necessary to understand the words of an Italian opera to know which singer's voice is superior. Songs, whether sung by humans or animals, do not have to contain information to strike emotional chords.

Scientists assume that female whales are judging the humpbacks' songs in the same way female midshipman fish judge their males, and female songbirds their prospective mates. Listeners to the humpbacks' performances may be able to make assessments about the size of a singer based on his pitch and ability to sing the song compared with other singers. The whale songs contain both lower and higher frequencies, graded sounds that move up and down the frequency scale, and changes in loudness. Some scientists speculate that the lower frequencies are directed at other males, perhaps as warnings, while the higher frequencies are sung to woo the females. The grades and modulations of frequencies may reveal something about the singer's "style" and thus his identity. Females likely recognize the voices of individual males and will choose the same male from year to year.

Marine scientists still debate the actual purpose of humpback songs, but based on more than 30 years of research, most agree that the songs serve several roles in humpback whale communication.

Scientists assume that males are the only ones that sing. Songs have been recorded as male humpbacks migrate between polar feeding grounds in the Arctic and Antarctic and tropical breeding grounds in the late winter and spring months. Most of the singing, however, takes place at the breeding grounds during the mating season. Scientists who study whale songs have focused most of their recent attention on the breeding grounds near the Hawaiian Islands, which are protected by the National Marine Sanctuary Program of the National Oceanic and Atmospheric Administration.

While the mating purpose of the songs seems straightforward, Tyack and others have found that males also react to each other's songs in a variety of ways. The songs may allow males to judge their distance from each other and avoid contact. Long-distance vocal calls on land also help species, including birds, canids (dogs, coyotes, and wolves), and primates, keep space between individuals. Humpback songs appear to contain features that identify individual singers, so males may listen in to each other's songs to assess their relative strengths and degrees of dominance. When a male finally finds and is accepted by a female, he will guard her to prevent other males from approaching. If surrounding males recognize that the successful male is older, larger, and more dominant, they usually swim away. Smaller male humpbacks will sometimes form coalitions and attempt to take a female away from a more dominant male.

Significantly, the whales' songs are not genetically programmed but learned. "Whale song can thus be seen as a form of cultural evolution, in the sense that the song is a learned sense which evolves," according to a report by Payne and Tyack, "Most changes do not occur between seasons. Instead, they occur during the time when the whales are singing, developing their songs methodically in measurable steps. Furthermore, the types of change varied from season to season, and so could not be attributed to repeating seasonal factors. We know of no other animal where whole populations introduce such complex, rapid and nonreversing changes into their vocal displays, abandoning old forms and replacing them with new."

Whale communication research has its origins in Cold War technology. U.S. Navy personnel discovered whale songs in the 1950s after the navy had deployed the then-top-secret underwater Sound Surveillance System, SOSUS, used for tracking submarines and ships in the Atlantic and Pacific oceans. The humpback's love songs recorded by SOSUS were initially classified by the government—not because there was anything secret about the songs but because the government wanted to keep the existence of SOSUS a secret. After SOSUS technology was made available on a limited basis at the end of the Cold War, Christopher Clark, director of the Bioacoustics Research Program at Cornell University, was given access to the underwater network of sensitive hydrophones to listen to the vocal signals of many whale species.

Currently, SOSUS is part of the navy's Integrated Undersea Surveillance System. Since 1993 the Bioacoustics Research Program has had access to parts of the IUSS network for whale studies. According to Clark, "In the first afternoon, with the help of Navy analysts, more blue whale sounds were detected than had been described in the entire collection of scientific papers. Within a day we had confirmed detections of blue, finback, humpback, and minke whales."

SOSUS can detect the calls of blue and finback whales as far away as 1,500 nautical miles (1,800 land miles). Humpback songs can be detected from about 300 hundred miles, and those of minkes from distances of up to 100 miles, Clark says.

Whale songs are often compared to birdsong, which is used for attracting mates and for territorial defense. Tyack, who specializes in whale and dolphin communication, demonstrated for me in his lab that when played at high speed, tape-recorded humpback whale songs sound identical to birdsong. Tyack says it's probably just a coincidence, but maybe it's not. It's a mystery.

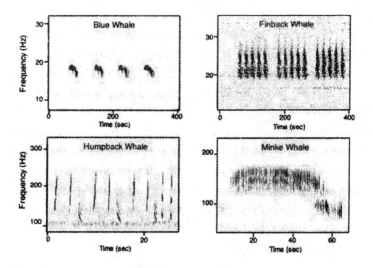

Spectrograms of animals' voices are a like a voiceprint. They reveal the structures of vocal calls. Comparisons can reveal important differences in the functions and types of calls.

The ocean presents a number of significant challenges for studying marine mammal communication and behavior in general. Scientists can't just go out on the ocean, or into it, as they do in the jungle and sit for days or weeks to observe behavior. First, they need vessels to get out to sea; they must contend with unpredictable weather and animals that are usually on the move. Second, marine mammals spend most of their time underwater. Scuba-diving equipment and diving skills or small submarines are needed to study the animals for any length of time. Boats are expensive, scuba diving in the open ocean is dangerous, and submersibles are hard to come by. Research on captive marine mammals is perhaps more practical, but it is limited to species that can fit into a pool and does not reveal as much about their natural behavior or the contexts in which they use their various signals.

Tyack has developed a recording device with suction cups that can be placed on the body of a whale when it surfaces. The digital acoustic tag sticks to the animal long enough to capture the sounds the whale makes during deep dives and to record information related to speed and depth that would otherwise have been impossible to gather. One of the species Tyack has been able to study is the endangered northern right whale, of which only 300 remain. (The right whale's name dates from the era of whaling on Cape Cod, when the right whale was considered to be the "right" one to kill for its oil.) "Our ignorance of the reproductive behavior and the mating system of right whales is so profound that we do not even know the season and location where mating takes place," Tyack says. "This hinders our ability to determine the effects of disturbances like vessels and to minimize them. We hope to provide data critical to understanding reproduction in ways that may point to how we can enhance the recovery of this most endangered of baleen whales." According to Tyack, collisions with vessels account for 35 percent of deaths of northern right whales. Tyack has found that right whales can hear and localize natural sounds very well, but they do not appear to understand or react to the sounds of oncoming vessels. His data show that right whales are remarkably buoyant and easily float to the surface from various depths, but they may lack control over their movements when surfacing for air. Learning what sounds the whales do respond to will enable vessel captains to warn right whales of a potentially harmful collision. "We need to address questions like what age and sex classes, behavioral contexts, or habitats pose the greatest risk," Tyack says. "What vessel characteristics, such as speed, maneuvering, and acoustic signature, pose the greatest risk? The tag will enable us to learn a lot more about how the whales respond to vessels as they get closer and may eventually help us develop an acoustic alerting system." The graph below is an example of the data being collected on the movements of a right whale by means of the digital acoustic tag.

Marine mammal researchers have become increasingly concerned about the effect of human activities on whale courting

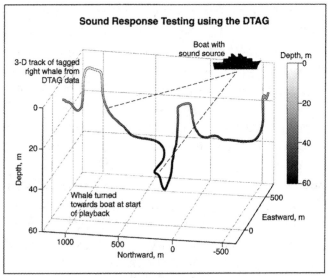

Sound Response Testing using the DTAG

SOURCE: Peter Tyack, Woods Hole Oceanographic Institution

The **digital acoustic tag** is able to track marine mammals through their normal daily routines, revealing important behavioral information on their communications and patterns of movement.

behavior, including the humpbacks' ability to get their songs heard by the intended receivers. Clark has studied the effects of man-made noise on humpbacks' songs to determine whether the sounds from oil and gas exploration and low frequency active sonar is interfering with the ability of humpbacks to communicate. Among other concerns is the navy's use of the natural sound channel that blue and finback whales have been using for long-distance communication for millions of years. "There are 100-year-old whales alive today who can probably remember when the ocean was a much quieter place, and they could communicate with colleagues across grand expanses of ocean," Clark says.

Clark has discovered that male finback whales produce very low frequency calls over long distances to theoretically attract females. Finbacks are leviathans, second in size only to blue whales, and can weigh up to 80 tons and grow to 85 feet in length. They do not appear to have any permanent breeding grounds like the humpbacks, so their strategy apparently is to congregate near the females' favorite dining places and sing to them, perhaps with the goal of mating. Unlike the tunes of humpbacks, finback songs are not likely to sell CDs, because they sound like pulses—boom, pause, boom, pause, boom. Nevertheless, the continuous, repeated nature of the vocalization qualifies it as whale song.

Clark conducts part of his research in the Sea of Cortez, where finbacks feed on swarms of krill, which are tiny, shrimplike crustaceans. Once the whales are together, the males are able to court. Some of the most intense singing occurs during prime feeding time. If a finback male is courting females at a krill swarm, man-made sounds could degrade its calls. Japanese fishing vessels also threaten the finback's ability to court females at dinner. Commercial fisheries have depleted populations of larger species, and Japanese vessels in particular are now taking krill, the primary food of large whales. Evidence is mounting that krill populations are drastically declining.

Some of the ocean's other singers include Antarctica's Weddell seals, which make trilling sounds during the breeding season, but scientists interpret these primarily as territorial and defense calls. Little is known about how females react to the calls or whether they react at all. Male walruses make a metallic bell sound during their mating seasons.

The largest amount of information on the songs of the animal kingdom comes from the migratory songbirds that provide the dawn chorus every spring in the United States, Canada, and Europe. Songbirds account for more than half of all birds on the planet.

Along with humpback whales, male songbirds are among the animal kingdom's top learners. The males of most North American and European species learn their songs, which are critical for attracting mates when they mature, from their fathers when they are chicks. They also learn the songs of their neighbors, which will be used in song-matching contests to defend territory. Learning songs is the most important thing a young male bird can do, and it appears to be similar to human learning. If his parents are a success-ful and attentive couple, and they are able to nurture him properly when he is developing, then the male bird can devote most of his energy to his studies. Research has shown that humans dream about things they have learned during the day, and that dreaming is a component of effective learning. A recent study led by Daniel Margoliash, of the University of Chicago, suggests that young zebra finches dream about the songs they have heard during the day. Margoliash and his colleagues discovered this by attaching small electrodes to four young zebra finches' heads to measure brain wave activity while the birds were practicing their singing during the day, and again while the birds were sleeping at night. The brain wave patterns during sleep matched the patterns recorded in the day when the birds were singing. To Margoliash, the finding suggests that birds dream about their songs and rein-force the neural circuitry.

Female songbirds carefully assess the songs of males of their species before choosing a partner and settling down on his territory. The cost of making an error in judgment can leave the female with-out a home and offspring, or with a male that is possibly the bird equivalent of a loser. The female songbird's success depends solely on her ability to select a healthy male that can control and success-fully defend a territory with plenty of resources. Before signaling the bird equivalent of "I do," the discerning female will investigate numerous male singers.

A study of black-capped chickadees by Tom Peake, of the Uni-versity of Copenhagen, and Daniel Lennill, of the Queen's Univer-sity in Kingston, Ontario, provides some insight into the female songbird's discerning nature. Male songbirds of most species arrive at their territories in late winter or early spring. The landed gentry from the previous season will reclaim their turf, while others fight among themselves for the best available trees and bushes. After the territories are sorted out and the females have selected their mates, neighboring males wage daily song-matching contests, mimicking each other and drawing from their early memorization of neighbors' songs. Some species learn the neighbors' songs after securing the territory.

Jack Bradbury says that based on studies of song sparrows by his wife, Sandra Vehrencamp, the song matching can be translated as something like "Yo mama." "No, yo mama." "Piss off." "No, you piss off." The birds usually drop the contest after a while and leave it at a draw. Sometimes one bird will become more aggressive and will continue to mimic his neighbor. In an all-out contest, whichever bird persists the longest or knows his neighbor's songs the best will win, which can have rather unpleasant results for the loser, especially if his mate is paying attention. And odds are she does not miss a note.

Peake and Lennill brought recordings of male black-capped chickadees to a forest to wage their own singing contests with the resident males. The scientists had the unfair advantage of battery power, which allowed them to win some of the contests. They discovered that the female partners of competing males do indeed pay very close attention to the contests. If a male loses, as one did during a singing bout against the recordings, its partner will soon mate clandestinely with the winning male. While songbirds are socially monogamous, they are not necessarily sexually monogamous. DNA studies show that up to a quarter of the offspring in a nest may be fathered by a neighbor. Apparently, the female loses confidence in the qualities of her defeated partner and mates with the neighbor to ensure that some of her offspring will possess winning genes. If a male loses two or more contests, the female will abandon him. Mating is a harsh game, especially for lackluster males.

The original discovery that songbirds are learners was made in the 1960s by Peter Marler, professor emeritus of neurobiology, physiology, and behavior at the University of California at Davis. Marler's work is among the most frequently cited by scientists in animal communication because he has made many key contributions. As a young scientist, Marler was studying birds in the Welsh countryside when he became intrigued by the observation that birds of the same species sang songs in slightly different ways at different locales. Marler coined the phrase "song dialects."

The presumption of scientists at the time was that birdsong was genetically programmed, as, they believed, were all other animal vocalizations. Marler reasoned that if birds of the same species were singing with dialects, they were capable of at least some degree of learning. Something other than rote programming had to be occurring. A series of classic studies soon proved this to be the case. Using white-crowned sparrows, scientists deafened young chicks and observed their song development. Normally, this species begins singing full adult song by about three months of age. A chick that was deafened before it reached the age of three months could only produce an abnormal song that sounded like a scratchy buzz. Species related to the white-crowned sparrow were shown to develop similar raspy songs if they, too, were deafened while still developing.

In a now classic paper titled "An Ethological Theory on the Origin of Vocal Learning," Marler reported that a male white-crowned sparrow that had not been deafened but was raised without being able to hear adult songs of his species also developed abnormal song. The song was not as abnormal as that of a deafened bird, and not as not normal as a bird allowed to hear the songs of adults while growing up. Marler concluded that the studies revealed a progressive loss of species-specific song—from early exposure to normal adult song, to rearing in isolation from adult song, to deafening during early development. Marler wrote: "A songbird, like a child, must learn from others if it is to vocalize normally."

The studies went on to show that if a white-crowned sparrow was raised without exposure to other members of its species but was able before the age of 50 days to listen to tape recordings of specific types of adult songs of its species, it would develop songs exactly like those heard on the tape. But if the young bird was exposed only to songs from another species, it could not learn those songs.

This means that the songbird is born with a species-specific genetic template of what its song should be, but the bird must be able to hear normal adults so that it can practice during a critical development period. The songbird requires the feedback of normal adult songs so it can compare the sounds it is making with those songs. With exposure to normal adult songs and a little practice, the songbird will develop normally and sing whatever songs it heard while growing up. The obvious analogy is to babbling in an infant, which requires a similar type of feedback to enable the infant to learn to use human language. Animal communication experts call the phases of a bird's song learning subsong, which is like babbling, and plastic song, which is akin to toddlers learning to use words.

A songbird's genetic template evolved so that the young animal will not be confused and mistakenly learn the songs of a different species that might be living within earshot. Nature is a very noisy place, with many species chattering at once. Having a genetic template on which experience can be layered is an evolutionary adaptation that reduces confusion in a world full of different chatty animals. The human brain also possesses a genetic template for learning and speaking human language. If the brain were truly a blank slate, a baby growing up in a house with a canary might accidentally learn to chirp instead of talk.

How long an animal is able to learn new sounds depends on the species. In most songbirds, the learning phase lasts only a couple of months, long enough for the birds to learn their repertoire of songs. These are the only songs the bird needs to defend a territory and attract a mate.

Scientists in England have become increasingly concerned about declining populations of finches, orioles, warblers, and other songbirds. They fear that the dawn chorus is endangered. The decline of songbirds in England is something of a mystery, but according to the Royal Society for the Protection of Birds, it may be due to the interference of traffic noise with the ability of young birds to learn their songs, particularly around London. Studies reveal that the songs young males are singing are substandard, reminiscent of the abnormal songs of the white-crowned sparrow experiments. The best these birds can muster are single-note chirps. Because males with substandard songs are unable to attract females as adults, the birds are simply not mating. Other bird species, including jays, are quickly filling the holes in the songbirds' previous territories.

Su Engstrand, of the Bird and Mammal Sound Communication Group at the University of St. Andrews in Scotland, has been investigating the effects of man-made noise on song learning in blackbirds, especially on song matching among neighbors. Blackbird song contains a relatively soft high-frequency twitter, which is being masked to some degree by background noise. Studies on the effects of noise pollution on songbirds are quite new and may help explain the cause or causes of dwindling populations.

According to Eugene Morton, only 13 percent of North American bird species defend territories yearlong. Real estate plays an interesting role in the types of vocal signals birds evolve. In the tropics, 65 percent of birds establish and defend permanent territories and develop permanent, monogamous bonds. While songbirds dominate as singers in temperate regions, both males and females sing in the tropics and often do so in dazzling duets. Duetting, as it is called, serves multiple purposes, not the least of which are maintaining the pair-bond and allowing the birds to keep tabs on each other's whereabouts when they forage out of sight of each other in the denser rain forests.

The yellow-naped Amazon parrot of Costa Rica is the largest parrot seen in the beautiful Guanacaste region the south, which contains large preserves. With time and patience, these parrots, typically seen in pairs or small family groups, are relatively easy to find. Males and females are the same size and have the same colorations and markings. The pairs tend not to be very social except with each other and their offspring. They forage in the day and reside at permanent roosts at night. At least three regional dialects have been reported in this species in Guanacaste. Pairs make soft contact calls when they are close, usually at their nighttime roost, but they have several other distinctive calls in addition to their duets, including loud contact calls, usually made by one member of the pair followed closely by the other. They also make a high-frequency preflight call just before and during their takeoffs. During aggressive encounters with other pairs, they squeal.

Jack Bradbury, who has been studying tropical parrots in Costa Rica, says, "The duets follow a common general syntax, with each member of a pair contributing according to general rules." The duet begins with both members alternating the loud contact call. After a number of these calls, the female begins making rough, raspy calls, also known as squeals. The male finishes the duet with several musical yodels. Bradbury says the yellow napes engage in duetting wherever they happen to be foraging each morning, whether other pairs are nearby or not: "If other pairs are sufficiently close, or when a pair is near to its nest, pairs may engage in counterduets in which each pair replies to the other's duet with its own version. Pair duets also exhibit dialect differences, with duet notes changing at the same boundaries as contact calls. The basic syntax of duets, however, seems to be preserved across dialects."

Duetting can serve as a primary territorial call, letting others know that the pair-bond is alive and well and the couple are working their turf. It is most common in tropical birds, but gibbons, the smallest of the apes, also engage in duets—which have been described as among the most beautiful and moving songs of the animal kingdom.

In the rain forests of Asia, pairs of male and female gibbons, which are among the 13 percent of primates known to be monogamous, sing elaborate songs that can last for half an hour or more. Loud territorial calls are common in most forest-dwelling primates for the same reason vocalizations are important in any species that lives in densely wooded areas: sound provides the best long-distance communication. Gibbons' duets convey possession of a territory and also let unmated males and females know that the singers are an established couple. Male and female white-cheeked gibbons sing different songs during their duet. In some gibbon species, the male initiates the songs, while in others the female sings first. Among white-cheeked gibbons, females initiate singing, which usually, but not always, prompts the male to join in. All but one species of gibbons alternate singing during the duets.

The **yellow-naped parrot** is able to mimic human speech. Mimicry among parrots of many species aids in social bonding. The ability of a new parrot to mimic a group's vocal "signature" may open the door to group acceptance.

Juvenile male gibbons sing a more femalelike song in duets with juvenile females. When the juveniles are singing at the same time as the adults, the added voices strengthen the message of territorial ownership for the gibbon family. As the males near adulthood, they sing a cross between a male and female song, which some studies suggest is a signal to Dad to kick Junior out of the group. By the time the young adult male is on his own, after a long learning curve of eight to ten years, he sings a strictly male version of the song by himself, which appears to announce to unmated females that he's available.

Gibbons' duetting may not involve the same type of learning by mimicry as in whales, songbirds, and parrots. It appears to be more genetically programmed. Pairs sing their duets primarily at dawn, but like the howlers, they also sing during the day in encounters with neighbors or territorial interlopers. Male and female gibbons are about the same size and seem to share equal status, which tends to be the case in most species in which males and females are the same size. In species in which males are bigger, males tend to be dominant. Among these same-sized gibbons, females are often dominant when it comes to getting and eating food and chasing away strangers, but during border skirmishes males do the chasing and fighting while females scream at the intruders. Sometimes whole families on both sides of the skirmishes square off along a border—adult males against other adult males, adult females against adult females, and younger offspring against younger off-spring. Unaware of the territorial aggression of the adults, young primates of different species commonly play together when two groups encounter each other. Duetting is characteristic of more so-cially complex species, as is the ability to learn songs, and appears to have evolved independently in primates and tropical songbirds as a successful way of communicating ownership in species that form monogamous relationships and defend territory.

Scientists have found few singers among mammals, although the largest of all, the African elephant, is at least technically a singer. Elephants are unique in the mating game because females do the advertising, calling to the males and letting them sort out who is dominant and wins the opportunity to mate.

Elephant societies are divided between females and males. Females live together in cooperative family groups headed by a dominant matriarch. Unless human predators kill them, elephants live to about 70 years of age and their sexual development parallels that of humans. Young bulls live at home until they become adolescents, when they run off to join bands of other single males. The typical family consists of the matriarch, her female offspring, and their offspring.

The matriarchs, like our grandmothers, possess valuable knowledge that they can communicate and that helps their families' odds of survival. In one study, Karen McComb, of Sussex University in the England, tested 21 elephant families with matriarchs that ranged in age from 27 to 67 years old. When matriarchs are killed, either by legal culling, illegal poaching, or angry farmers, the effect is as profound as the death of a strong, wise grandmother in a human family. Knowledge of watering holes that are reliable in times of drought is lost. Familiarity with the calls of other elephant groups is lost, which makes it more difficult to avoid conflicts between herds. When dominance hierarchies change, a family might lose a food supply that the matriarch had commanded for decades. Elephant families that have lost their matriarch may grieve for years and sometimes drift apart.

Like dialects in songbirds, contact calls differ from one elephant family to another. Spectrograms of different families' calls reveal differences in the patterns of the calls and features unique to each family member. To learn how well one family recognized the contact calls of another family, McComb played recordings of the contact calls of 21 families to each family separately. The oldest matriarchs were much better than younger matriarchs and younger family members at recognizing other families' contact calls. When a matriarch hears a call that is unfamiliar, she will signal to her family to group together, a defensive move to protect the family's calves, because sometimes encounters with other families can become hostile. McComb's study documented that older matriarchs were best able to recognize unfamiliar calls and gather the family. Younger matriarchs did not display the same degree of caution or savvy.

Mating and birth are seemingly sacred occasions in female elephant society. Scientists who have studied elephants in the wild for decades have said that families appear to rejoice after a female has mated and when calves are born. McComb found that older matriarchs were more effective at ensuring that females in estrus leave the group to make their long-distant mating calls. Her studies found that females in families with older matriarchs have more calves than females in families led by younger matriarchs. Grandma is a little wiser than younger moms and is vital to the stability of the family.

At about the age of 20, females are generally mature enough to start finding mates. They become sexually receptive for only a few days every four years or so, a small window of opportunity for both females and males. Meanwhile, at about age 10 young males run around for up to a decade before they mate, their raging hormones making them resemble teenage boys. They appear to be obsessed with sex but are rarely a match for the old bulls and must wait their turn. Young elephant bulls can be unpredictable and mischievous and have been known to form coalitions to corral a female.

When the time comes to mate, the female in estrus leaves the family temporarily to begin calling for her male suitors. Once she is alone she emits a series of low-frequency sounds, or infrasound, produced mostly below the level of human hearing. These are the most intense, continuous calls that a female makes during her lifetime, and, to scientists, they qualify as song because of their continuous, repetitive nature. Under the right atmospheric conditions, a mating song will radiate out as far as six miles.

The ideal time of day to produce these siren calls is late afternoon, when the temperature cools and begins pushing warmer air upward. By dusk, a ceiling of warm air will have formed about 300 feet above the ground. This creates a sound channel, an invisible tunnel similar to the sound channel in the deep ocean used by blue and finback whales for their long-distance calls. Low-frequency sound waves normally have high peaks and valleys, but the channel keeps the sound concentrated and forces it to travel much greater distances. (Animals, it would seem, were the first to learn the value of evening rates for long-distance calls.) A midday call would travel only about two miles. The female that makes her long-distance call in the evening gets greater distance for the same expenditure of energy. With her long-distance, omnidirectional call, the female can cover sixty square miles during optimal evening conditions.

The signaling elephant pushes air from her massive lungs through the trunk, which is ringed with layers of muscles and bundles of nerves that provide remarkable dexterity. The elephant is able to produce higher notes and frequencies by constricting the muscles in her trunk, which is an amazing natural musical instrument. When an elephant is generating low-frequency sound, the skin of her forehead puffs out and vibrates, a bit like Louis Armstrong's jowls.

After a female makes her love call and the bulls show up, the males joust for dominance by pushing against each other with their foreheads, although they rarely injure each other. The fighting is usually nonviolent posturing unless two bulls are equal in status and size. Strength, size, and determination win. If the female has mated before with a dominant elephant that she recognizes, she will often prefer him again. The successful male will attempt to guard his female companion until she has passed through estrus and is no longer sexually receptive, but female elephants sometimes have other plans. When the bull leaves, the female sometimes calls again for another partner.

When the mating is complete, the female returns to her family and is greeted by her sisters, aunts, and mother with a noisy, excited welcome-home party called the mating pandemonium. All the females scream, trumpet, urinate, defecate, and release secretions from their temporal glands. Similar postcopulation behavior can be found in some human cultures in which excited villagers greet a newly married couple the morning after the nuptial night.

Most of what has been learned about elephant communication can be credited to Joyce Poole and Cynthia Moss, who worked together for more than two decades at the Amboseli Elephant

Research Project in Kenya. In one of their classic studies, they found that only 30 percent of more than 1,000 hours of recorded elephant calls were audible to the human ear. Over the years, Poole and Moss have learned to recognize specific meanings of more than 30 different elephant calls, which they group into nine call types.

The greeting rumble is used between elephants after they've been separated for a while. Family groups associate with as many as five other families, called bond groups, which are probably related. Poole and Moss reported as common behavior a story about a reunion between two elephants, Fernanda and Flavia, that belonged to different families in one of these bond groups. Fernanda had been feeding among several hundred elephants when Flavia arrived at the group. At about 60 feet away, Flavia began to rumble at Fernanda, who recognized the call and responded immediately, according to Poole and Moss. Greeting rumbles increase in intensity and volume the longer two elephants have been separated. Fernanda rushed out from her family group with her two-year-old in tow and began trumpeting very loudly. The calf, imitating Mom, also started trumpeting. Temporal gland secretions, looking quite like tears, were streaming down Fernanda's face as she approached Flavia. The two kept rumbling and trumpeting, ears raised and folded, as they met. Their excited greeting was accompanied by a fair amount of urinating and defecating.

By contrast, the contact call used to gather members of a family or a bond group sounds rather soft. The elephant that makes the call flaps her ears as she vocalizes. After issuing the call she raises her head, spreads her ears, and listens for a contact reply. The receivers of the call raise their trunks to blast a sound the moment they hear the call. Poole and Moss recorded a female and her family making contact calls and replies for several hours when they separated more than a mile from each other. Members of a bond group clearly recognize each other's voices.

When a family member wants the group to move on after

they've been grazing or hanging out at a mud hole, she makes the "Let's go" rumble, as if she's getting the family together to leave the house for a trip or to go out to dinner. The caller walks to the edge of the group and, while facing away, lifts one leg, flaps her ears, and makes a long soft vocalization. One by one, the elephants begin to move. Poole and Moss watched an old matriarch named Wart Ear make the let's-go rumble. First one female started to move, then, after a while, another. A bit later, another got going, until finally the whole group was responding. Any males that have been associating with a bond group ignore the command.

The musth rumble is the "Baby, I need you" call of the male when he falls under the powerful influence of sex hormones. The call sounds like a low-frequency pulse. As the male calls, he flaps his ears, then stands still with his ears flared as if listening for a response. Other musth males that hear the rumble take notice of its direction so as to avoid bumping into the caller, which is sure to lead to a fight.

The female chorus is the response to the musth rumble, but it doesn't always result in mating because the females of a family group might not be in estrus. The exchange does result, however, in the males and the females gathering to investigate each other by testing urine for olfactory signals and such. The loud calls of the females might also let other males know their location.

The postcopulatory sequence is a repeated low-frequency call made by a female after she has mated, which seems to elicit guarding behavior from the male she is with while also attracting other males. She often makes this call for half an hour, then makes it again. A study of a female named Zita revealed that after she mated with male 10, he guarded her for two days. After he left, she mated with male 22, who also guarded her for a few hours before losing interest. Then she mated with male 7, who had been waiting at a safe distance.

Both males and females make rumbles, trumpeting sounds, grunts, screams, purrs, and infrasound. When families are moving through the bush within earshot, they make audible grunts as contact calls. If one of them spots a human or other danger, they become silent. If they are spread apart over a few miles, they use infrasound to coordinate their movements. Smell and hearing are the most important senses in elephant communication, but they also make visual signals with the ears and trunk. An elephant that is flaring its ears and shaking its trunk is angry. If it begins trumpeting in a harsh manner, it is ready to charge.

Often when an elephant charges, it will stop at the last minute, which had been the experience of Mike Fay, a conservation biologist with the Wildlife Conservation Society. Fay has spent more than two decades in Africa and trekked 2,000 miles over 14 months through Congo and Gabon. He had been charged and stood his ground, but on New Year's Eve 2002, he encountered a hostile elephant that was not bluffing. A young female accompanied by an older female and a calf gave the warning signs and then charged, pinning Fay to the ground and puncturing his right biceps with a tusk.

Over the past decade, scientists have begun to learn even more intriguing aspects of the silent communication of elephants. Caitlin O'Connell-Rodwell, who began studying elephant communication in the 1990s, observed that elephants will raise their feet sometimes in a peculiar manner and also bend forward to stand on their toes. Elephants have been observed gingerly touching a dead calf with their toes as if they were using their feet as a sensory organ, which they are. Their feet are extremely sensitive to vibrations. They may touch a dead calf to sense whether it is breathing or has a heartbeat.

Elephants are also known to have a mysterious ability to sense thunderstorms from long distances and begin moving toward them for the promise of water and a relaxing mud bath. On these occasions they raise their feet and bend forward just as O'Connell-Rodwell described, apparently feeling through the ground the vibrations of the thunderclaps generated by lightning strikes. In the late 1990s, O'Connell-Rodwell conducted playback studies during which she sent signals through the ground to captive elephants. Most of the time, the elephants turned in her direction after the signal had been sent. O'Connell-Rodwell also sent a series of different calls, including warnings, and found that one older female became so agitated by the signal that she knelt down and bit the ground, a behavior also observed in the wild.

Elephants can sense the direction of a sound source based on which feet the vibrations reach first. If an elephant is facing north and the vibrations reach its right feet first, it knows that the source of the sound, such as a rainstorm, is coming from the east. The elephant raises the right front leg and then the left to determine which foot the vibrations reach first. The elephant's toenails conduct the sound into its legs and up to its brain through nerve or bone conduction. Elephants have sphincter muscles in their ears that allow them to close the ear canal off from sound.

The maximum distance over which an elephant can detect seismic vibrations is about 20 miles. O'Connell-Rodwell says the infrasound calls used by elephant groups may propagate through the ground much farther than through the air. This would make it possible for one group to detect infrasound signals from another group and judge its distance and even its direction of travel. The didgeridoo, a hollow, trunklike instrument used by the Aborigines of Australia, can send seismic vibrations through the ground. O'Connell-Rodwell's group has begun investigating a wide variety of animals that may augment communication via ground signals, including lions, which roar while lying down and may detect the vibrations of running prey.

The scientists of song and sound have been able to learn much about the ways that marine mammals, birds, and even elephants attract mates and in some cases employ their signals to guard their turf. But animals have developed many other colorful and flashy ways of attracting members of the opposite sex. Let's join them now and, in the words of David Bowie, "Let's dance."

Seven

FLASH AND DANCE

H UMANS DID NOT invent dancing, sexy attire, or romantic gifts. As members of the animal kingdom and speakers of the natural language, we do have the same urges as others to sway, spin, and shuffle our feet, to dress to impress, and to entice each other with flowers, sweets, and sparkly stones. These behaviors are as innate as growling, singing, and discovering that the smell of another person's skin is powerfully seductive.

Stimulating the visual senses to attract another animal's attention tends to bring out the best in a species' creativity. In this regard, nature is unbounded. To the uninitiated, the courting rituals of animals may appear strange, especially the more energetic routines in which the members of the opposite sex often engage when attempting to attract a mate. But dance is an interpretive art. A few years ago, I happened to be in a snow-covered village in Nepal called Gorak Shep, which translates loosely to "graveyard of the crows." Situated at an altitude of more than 16,000 feet, it is the last stop before climbers reach the base camp of Mount Everest. How the village got its name is a mystery; there were plenty of crows hanging about. (If there were a real yeti, the crows probably would have led it to us.) I was covering an expedition with members of the Explorers Club and was nine days into a ten-day trek to base camp. We had awakened to a brilliant sunrise and pure blue skies. My gear was packed and I was putting on gaiters, which fit around one's calves and boots to keep the snow out. For some reason the gaiters made me think of the leg warmers popular in the 1980s, particularly the ones featured in that particularly awful movie *Flashdance*. At that moment, in front my tent, the Nepalese porters were lining up 30 yaks, on which they began loading our gear for the push to 17,500 feet. The next thing you know, I was on my feet doing the running-in-place flashdance thing in front of the yak that had been carrying my gear, and singing loudly, "She's a mani-yak, mani-yak on the trail. And for nine days I've been fol-low-ing her tail. Mani-yak, mani-yak . . ." The guys who knew the movie got the joke, bad as it was. But the porters, who had no context for my display, only saw an American behaving very oddly; some of them looked a bit frightened. At least the yak showed no signs of misinterpreting the number as courtship.

Seen in their natural context, the exuberant courtship displays of mammals and birds are truly impressive and fairly easy to interpret. Native American tribes that once lived on the western plains of North America were so inspired by the courtship rituals of the male sage grouse that they interpreted some of its dance moves for their own rituals.

When it comes to visual displays, there is only one basic rule: If you've got it, flaunt it. In the wild, it is usually the male who dresses up, adorns himself, and dances for sex. Because of the demands of females over the past 500 million years, visual courtship displays have come to honestly represent the vitality of the male. Hands down, visual signals are the most expensive to generate for communication, demanding a great expenditure of energetic resources.

Before moving to the dance floor, let's look first at how some species have dressed and ornamented themselves to impress potential mates. Patterns on coats, feathers, and scales have remained popular for millions of years and show no signs of going out of style, with the exception of a few species that prefer to molt and generate a new skin from year to year. Body stripes are common among mammals, fish, and some birds for accentuating physique. Stripes that make easy comparisons of male body size possible at a glance are most useful for females evaluating a potential mate. They also make it easy for competing males to size each other up. Contrasting light and dark facial markings are certainly attractive, but they are most practical for allowing recognition of a species at a distance. Before signaling, one needs certainty that a potential mate belongs to one's own species. Often a male's markings will highlight some feature that is highly desirable to the female. Outlines such as those of the monarch butterfly are useful for enhancing shape and visibility against the backdrop of the sky. For species with ultraviolet vision, iridescent attire can generate brilliant colors with the proper lighting. The red-tailed tropicbird creates contrast with its snowy white body and bright red tail and beak, which stand out nicely against the blue skies above the Pacific Ocean. Finally, it is truly a fashion faux pas for any species other than the original owner to wear fur.

Birds are among the animal kingdom's more frequent flashers. Some species hide colors beneath feathers for flashing during courtship or to draw attention to a decorative ornament, which males of some species have developed to an exaggerated degree to impress females. The rooster's comb is as sexy a feature to hens as the peacock's long train is to peahens. Ornaments are technically defined as "visually transmitted physical traits such as color patches, feather plumes, elongated tails and fins and any other non-weapon body structure." Weapons are things like antlers. Even though antlers usually grow for the mating season and then fall off, their purpose is to intimidate other males and to win head-on battles for dominance. Females do not ordinarily adorn themselves with ornaments, but the highly visible swollen red bottoms of some primates, including female chimpanzees and baboons, fit the definition.

Scientists have developed different ideas to explain how long tails, head appendages, and other ornaments evolve. The most popular explanations are the Fisherian runaway model (named for ethologist Ronald A. Fisher in 1930), the good genes model, and the direct benefits model. The Fisherian model pertains to male traits that are arbitrary to begin with—that is, they are not tied to any indicator of fitness in the male. According to this model, the female preference for the trait, such as a slightly longer tail feather, and the male trait evolve together. The more the female prefers the trait, the more it is selected genetically. The genes in the female that are linked to her preference, and the genes in the male that are linked to the trait, are inherited by their offspring. The process is self-reinforcing and self-perpetuating.

The elaborate feathers of birds of paradise appear to be an example of the Fisherian runaway model, according to Bradbury. Each genus among birds of paradise has developed different and arbitrary ornaments. One genus has elongated tails, another has head plumes, another has long filamentous body feathers, and yet another possesses curly tail feathers. One of the more unusual traits that may be an example of the runaway model can be found on a South American rodent called *Agouti paca*. According to Bradbury: "The penis of some rodents, such as the South American *Agouti paca,* is covered with spines. It seems unlikely that the structure provides direct information about male quality . . . The movement of the penis in the vagina cannot go unnoticed by the female, and suggests that it produces a tactile signal that is preferred by females."

The good genes model proposes that the ornament began as a trait that was a direct and honest indicator of the male's fitness, which corresponds with Amotz Zahavi's handicap principle. A male trait that is costly, such as the peacock's tail, or heavy antlers on a male elk, is preferred by females and thus selected and passed along to future generations. The direct benefits model is represented by the male fire-colored beetle and his nuptial gift of chemoprotective blister beetle juice to his potential mate. In general, exaggerated ornaments are expensive, as is a Hummer vehicle.

Among the least expensive and most effective signals is a simple badge of contrasting color that provides a female with immediate information about a male's social status. The dark badge on the male house sparrow's lighter brown chest grows in size with age to reflect his increasing social status. A female house sparrow selects a male with a larger badge because it tells her that he is a survivor who has won contests for dominance and possesses prime real estate. The badge, in fact, is linked to higher levels of testosterone, which is tied to dominance.

Scientists tested the importance of the badge as a signal of dominance in a classic study in which several sparrows suddenly found themselves living the high life, only to be punished later for a crime they did not knowingly commit. The scientists first captured several young subordinate sparrows, painted badges of dominance on their breasts, and released them. Early on, the other sparrows treated the imposters as if they were generals. The imposters were allowed to eat first and were accorded all the privileges of rank, but it didn't take long for the others to catch on to the ruse. For one thing, the subordinates would have lacked the unmistakable scent of dominance. When the game was up, the genuinely dominant males attacked the young sparrows without mercy. In nature, deception rarely goes unpunished.

The males of one variety of great tit birds, of which there are about 200 species, display a black stripe on their chests. These birds are monogamous, and wider stripes are associated with better parental care and dominance. Females, naturally enough, select the males with the widest stripes. Recent research in male lions suggests that the color of the mane is a badge of dominance: the darker the mane, the more attractive the lion is to females. The darker color also appears to be tied to testosterone production. In lion society, males and females do not form pair-bonds, and females as a rule take charge of mate selection, but older, experienced males sometimes have their own preferences, as do dominant male chimpanzees, and ignore certain female suitors while favoring others. Animals are born with different temperaments, something akin to human personality. Perhaps they prefer the companionship of other animals with particular temperaments. A safari guide in Africa described an occasion when he saw several female lions trying to stimulate a dominant male to mate. The male rebuffed the advances of these females and appeared to be interested in a single female that, alas, was not interested in him.

Bright colors are the fashion statement of the males of many bird species because they are an important sign that the males have "good genes" for females selecting mates. Among European and North American songbirds, males tend to be the ones with the brightly colored feathers—except for tropical species, in which colors are important to males *and* females for camouflage, which is a plus for mothers because it makes them less conspicuous to predators.

Brightly colored feathers, like nicely tailored worsted suits, are costly. The colorful male invests significant nutritional resources in his plumage. To the female, the colors may be pretty, but more important, they reflect whether the male has been maintaining a healthy diet. A good diet would mean he must have a territory with good food resources to provide his pigments. The idea is no different from humans knowing from his appearance that a well-groomed, well-dressed man with an athletic physique takes good care of himself.

Geoffrey Hill of Auburn University studies what certain colors may be communicating to females. His group's research on North American house finches found that the diets of the males provide them with carotenoids, which their bodies synthesize into the species' characteristic bright red or yellow feathers, just as excessive carrot consumption can turn a human's skin orange. Hill's studies confirmed that female house finches prefer males with the brightest colors. Studies have shown that this is also true in tropical fish that obtain pigments for their bright colors from their diets. *Pseudotropheus demasoni,* which is bright blue with dark stripes, derives its coloration from eating spirulina algae. Many pet stores sell fish food with carotenoid supplements designed for specific species to help owners maintain the natural, colorful beauty of the fish.

Male lazuli bunting finches are known for their bright blue plumage, which can, however, vary from bright blue to a dull brown. Female buntings are of course attracted to the males with the brightest blues, which develop when the male is mature. This species has worked out an unusual alternative to the colorfast strategy, so that both pretty boys and dull boys get more than their fair share of mating. This is what scientists call disruptive selection, in which nature favors extreme traits at both ends of the spectrum at the expense of moderate ones. The dominant males, which have the brightest feathers, have their pick of territories and are most successful at attracting females, which are predominately brown. Erick Greene and Bruce Lyon, of the University of Montana in Missoula, were surprised to discover that the dullest males in a lazuli population were mating successfully and

Adult male lazuli buntings develop bright plumage, which attracts females and conveys information about their high social status. The brightest fellows get the best territory, but they will invite males with dull plumage to move into the upscale neighborhood in exchange for mating with their wives.

living unchallenged in some of the best shrubs, as close neighbors with the hotshots. Meanwhile, the hotshots forced moderately colored males to settle for scrubby bushes, where they had a tougher time attracting mates.

Imagine a college fraternity house in which the biggest nerds on campus are living side by side with the richest jocks, and at the jocks' invitation. The condition under which disruptive selection arose in the buntings was probably a shortage of good housing. The area where the odd arrangement was observed was a patchy habitat with fewer available shrubs than normal. Such conditions would intensify the competition for habitat and females. But why would the hotshots want to share with the dull boys? The answer appears downright Machiavellian.

Equipped with a nice shrub, the dull boys easily attracted mates. We know that human dull boys with significant economic resources have no difficulty attracting mates either. So rather than sharing the neighborhood with average-looking guys who would be tougher competition, the hotshots let the dullards move in to get access to female neighbors who might be more easily seduced.

A yearling male lazuli bunting is still a dull brown, but it will soon acquire the species' characteristically bright blue feathers. Some adults are brighter than others. The most brilliantly colored males are favored by females. The colors, as they do in many males, represent a signal that says "I am healthy and able to acquire and defend territory."

The species is socially monogamous, which means that females select one nest mate. But social monogamy is not sexual monogamy. DNA testing found that half the eggs in the nests of the dull dads belonged to the hotshots. So even though the dull boys are allowed to move in and attract mates with their good turf, the females cannot resist the hotshots once they settle into the neighborhood. The females want those hotshot genes to pass along to their offspring. But everybody wins. Dull males, which would probably never attract a female, get a female, a nice home, plenty of food, and some offspring. The beautifully colored hotshots, for their apparent generosity, spread their genes throughout the community.

When it comes to dance, an animal does not simply wake up one morning and decide that it is going to try some new moves. Its routine has developed over time, under the influence of sexual selection, from movements that it commonly makes in its daily life. Intention movements, which have many communicative roles, have provided much fodder for the dances of the animal kingdom.

Flocking birds, which are pros at group decision making, coordinate their takeoffs with intention movements similar to the movements they would go through if they were about to fly. When a flock takes off, the birds appear to be remarkably in sync. The signal is similar to the quorum-sensing signal of bacteria in the way it spreads through the community. When one bird decides that it is time for the flock to move, it will raise its wings and maybe lift off a bit. The birds immediately around it catch on, and they too start making the signal. The signal spreads through the flock, and when a quorum is reached, the birds all lift off at once. Often, they are also vocalizing with a bird version of the elephants let's-go rumble. Among European blackbirds, the signaler crouches, lifts its tail, and leans its head back as if about to spring into the air. The signaler sometimes does not get a quorum, and no one leaves. Sometimes a portion of the flock may have been getting ready to take off, but they quickly settle back down when they realize the majority is not following. The geese that visit the Potomac River go through a similar consensus building by honking and jerking their heads before taking off. Just like the flight intention signal, positions, postures, exposure of particular body parts, and any movements an animal would typically engage in when preparing to copulate can become incorporated into signals associated with mate attraction. Over time, the movements become ritualized into the animal's behavior as a signal.

Insects have incorporated some curious intention movements into their courting rituals. Offering gifts is a polite custom in human courtship and may signify that the giver has more than just a passing interest in the relationship. Balloon flies also present gifts. The male balloon fly offers an empty ball of silk to the female of its species. This ritual probably evolved from an original present of the equivalent of a box of chocolates—a ball of silk containing a tasty prey item. Over time, males apparently "realized" that they could get away with eating a little piece of the prey item before wrapping it up as a present. Later, the males began wrapping up only a dried bit of prey that they might have found lying around. Eventually the males skipped the contents altogether and began presenting an empty package as a gesture.

The male wolf spider, in the truest sense of an intention movement, has taken a more direct approach. On its body are two long leglike structures called pedipalps that carry its sperm in sacs on their ends. During copulation it releases the sperm from the sacs. The spider's courting method is to approach a female and, while facing her, wave the pedipalps in circles as he would if he were about to mate.

When a male mammal begins his courting ritual, he will direct his gaze toward the female and then begin displaying in a species-specific manner. His objective is to close the distance between himself and the female as quickly as possible to make physical contact. The female will either run away if she is not interested or convey her willingness to mate with intention movements of her own. Typically, she will stand still and maybe nuzzle the male, or she may assume the position for mounting. Wild mares can easily run away from stallions that they consider inappropriate partners, but when they approve they stand perfectly still.

The degree of involvement of the female in the dance depends on the type of social relationship her species has developed. The juvenile albatrosses truly tango, both partners playing equal roles, because monogamous relationships such as those found in the albatross, tend to invite greater participation from females in the displays. In nonmonogamous—polygynous—species, the pressure is on the males to flashdance. Generally speaking, polygynous males have few responsibilities outside of eating, frequent mating, and fighting. There are, of course, exceptions to every rule in nature. Many male fish and amphibians take responsibility for the fertilized eggs of their frequent partners. Some males lose their heads during courtship—the female praying mantis bites the male's head off before mating, which prompts his sperm to be released.

Consistent with the universality of vocal signals, the creative, varied dances taking place in the wild rely on a few common steps on which a species adds its personal touch. According to James W. Davis, of Ohio State University, and Whitman A. Richards, of the Massachusetts Institute of Technology, the common movements found in so many different routines are the up-and-down, side-to-side, and circle motions. The albatrosses at Midway are head bobbers extraordinaire as are many species of waterbirds, including mallard ducks.

The key that makes any of these basic moves attractive is rhythmic repetition. A common come-on in lizards is the repetitive push-up, a version of the up-and-down movement. Falcons and red-tailed tropicbirds are among the species that have adopted circular flight movements. Hummingbirds make U-shaped flight movements similar to a pendulum swinging back and forth. Chimpanzees sway from side to side both for courtship and as a threat signal, recalling the swaying movements of a high school slow dance and those of a boxing ring. Davis and Richards classify all of these as oscillatory movements. The more complex the movement, such as circular flight and the U-shaped signal of the humming-bird, the more rarely it will be used. The simpler the movement, the more it will spread into the dances of all creatures from insects to elephants, which also sway.

Davis and Richards write: "It is interesting that certain types of oscillatory motion patterns appear across such a variety of species ... As for humans, some of our most basic forms of communication and expression include these same motion patterns, such as the simple repetitive nodding and swaying of the head and hands as responsive gestures."

It is all part of the natural language. Scientists would call it convergence.

Roadrunners fall into the up-and-down group of displayers and have adapted a version of the sway. These flightless birds are native to the American Southwest—the central valley of California, southern Texas, and northern New Mexico. Roadrunners gained Hollywood fame in Warner Brothers cartoons that featured a never-ending predator-prey relationship in which Wile E. Coyote was usually smashed by falling rocks or blown up with Acme dynamite. The real roadrunner, which is about two feet long and a member of the cuckoo family, got its name from settlers on their way to California. Like a dog chasing a car, the roadrunner ran alongside or in front of covered wagons as they traveled across the desert in the 1800s. With its long and powerful legs, the bird can run at speeds up to 17 miles an hour. The roadrunner is not especially attractive. It is brown and mottled to blend with the desert background, making life harder for predatory hawks, raccoons, and coyotes. The roadrunner itself is a ferocious predator—it catches its victim by the head, and holding tight with its jaws, it thrashes the prey back and forth, usually against a rock. Then it crushes the prey's head and swallows the battered meal whole. The roadrunner has been described as looking like a feathered dinosaur.

Early spring is the time for roadrunner courtship. Since the male has no astonishing physical ornaments for impressing females, he courts with a nuptial gift, which he carries in his bill and presents to a potential mate. He may offer a lizard, a rattlesnake, a field mouse, a bat, or even a scorpion, all favorite foods of the roadrunner. To approach a female, he employs stealth and races up behind her, gift in bill, and starts wagging his longish tail from side to side and alternately bowing to her. If the female is being coy, he may begin springing up and down, attempting to hover beside her. If the female does not want him, she refuses the gift and dashes away, but when sufficiently impressed with his performance she will accept the gift and mate. Roadrunners usually bond for life, building a nest together or taking one from another species, often in a hole in a tree. Both tend their eggs and hatchlings. They vocalize to each other using a series of six to eight dovelike coos that tend to drop in pitch as the series progresses. When aggressive, they can make a low-frequency clattering sound by grinding their jaws together. The roadrunner is very sensitive to cold, not unlike a reptile. To overcome a drop in body temperature during the cold nights of the desert, it has evolved a black patch of skin at the nape of its neck that is, for all practical purposes, a solar panel. In the mornings, roadrunners face away from the sun and lift their feathers to expose the patch. They stand there as long as it takes to warm up, with the patch rapidly absorbing solar energy.

The male Jackson's widowbird, native to Kenya, is strictly an up-and-down romancer. It makes its home in tall grass and creates a flattened patch of grass resembling a miniature crop circle as a staging ground for its dance number. To attract the attention of females, it leaps repeatedly into the air like an avian Baryshnikov and vocalizes loudly. "Hey, look at me! Here I am." Up it goes. "Hey, look at me! Here I am!" Up it goes again. The male leaves a thick pole of grass standing at the center of its stage and hides there from predators between displays while making soft cooing sounds until found by a female. Often, dozens of males may be displaying at the same time, which makes for quite a sight. The birds are dark-feathered for contrast against the tall yellow grass. When they leap, they tuck their tail feathers over their backs and resemble a feathery black ball.

The red-tailed tropicbirds on Midway Island perform a more complicated and rare maneuver above their nest sites—the gravity-defying circular aerial dance. The males arrive on Midway in early January and excavate a shallow nest in the low vegetation near the beach. Any vegetation will do as long as it provides shade for the future chicks. The wind is almost always blowing across Midway, in a fairly constant direction for about half the year before it switches directions. As a male flies directly over its nest, it faces the wind and hovers. The wind carries it backward in a sweeping upward arc until it reaches the top of its circle, then it drops forward to hover for a moment over the nest before repeating the movement. All the while, the male unleashes a guttural squawk to ensure that a female is watching his acrobatics.

As numerous males simultaneously perform their aerial displays, female red-tailed tropicbirds are drifting on air currents, watching the show. When they choose a male, they fly into his airspace, and both drop to the ground near the nest. The species gets its name from the male's long, slender red tail feathers, called vanes, which appear to serve a dual purpose: they provide stability for the hovering maneuver, like the tail of a kite, and they attract a female by serving as a colorful ornament. The males and females are about the same size and look nearly identical.

Great crested grebes have adopted the sway and the bob for a display called the weed dance, which begins when both the male and female dive underwater and snatch some grass in their bills, a movement associated with nest building. Next they paddle like mad to rise gracefully up from the water, breast to breast, and bob their long necks back and forth like Europeans kissing cheeks. The entire ritual of the grebe, like that of the albatross, is elaborate and lengthy.

Some of the more dramatic, energetic performers are polygynous birds, like the sage grouse, that display on a patch of well-defended ground known as a lek. Scientists have various theories about how and why lekking males go about choosing their stages. The hot-spot theory, developed by Jack Bradbury, suggests that males set up stages where they know females are most likely to be walking by. When people go hiking in remote places and find trails, we tend to assume that they were made by other humans, but animals create their own highway systems, and female grouse are no exception. The hot spot usually ends up being at the center of the congregated displayers. It is the grouse's version of center stage. When mating season begins, dozens of males position themselves at their leks, which are located in the same area year after year. The sage grouse's habitat is a flat prairie, but nonetheless, dominant males claim as their territory any bump that is a little higher than the others for their stage. The performance begins with an extremely loud series of vocalizations that are commonly described as coos, pops, and whistles. These "prairie chickens" use the vocal signals to help females locate them. The male sage grouse has evolved a pair of bright yellow inflatable air sacs on its breast to make itself more conspicuous. The male grouse's booming pop can be heard up to two miles away. To counter the risk of being heard by a predator, the sage grouse has developed a means of varying its signals in a sophisticated way. For the long-distance invitation, it focuses the sound waves into a beam.

Studies conducted by Marc Dantzker, of the University of California, have shown that the grouse's vocalizations travel anywhere from 40 percent to 100 percent farther than omnidirectional mate attraction calls like those of songbirds. The sage grouse's three types of calls are projected at different distances and in different directions. The pop beam travels the farthest distance and advertises the male's readiness to mate. The coos, which are the most intense, spread out in front of the bird as a sexier signal intended to woo the female closer. The whistles are directed out to the sides and may be intended to keep competing males at a distance. No other animal is known to create such asymmetric patterns of calls on land. During all these vocalizations, the male grouse beats its stiff feathers to create loud swishing sounds. Meanwhile it is strutting around like Travolta at the center of the disco floor. The males put intense physical effort into their visual and vocal displays as the females wander about.

Studies conducted by Bradbury and Vehrencamp indicate that the females prefer males that put the most energy into their dances. Birds that are not in tip-top shape will lose weight during their displays. The fittest maintain their body weight and are able to dance the most energetically and for longer durations. Females want their males to be strong and to demonstrate that they have superior endurance before they agree to mate. The female sage grouse wants only the best male genes to ensure a healthy genetic inheritance for many future generations.

The lekking strategy, which showcases the animal kingdom's dancing skills onstage, is widespread. Ducks, birds of paradise, whydahs and widowbirds, peacocks, sandpipers, snipes, bustards, and all the galliform birds, which include sage grouse and chickens, have developed amazing strutting displays that they perform on their stages.

The lek for birds of paradise species is usually a tree branch, which males fight over to gain the best position. The polygynous species wear the flashiest plumage in the animal kingdom. The male *raggiana* bird of paradise vocalizes to advertise its location to females and waits for a curious prospect to land on his perch. During his display he turns nearly upside down on the perch so that his tail is high and visible; then he spreads his wings and clasps them overhead to provide his guest with the fullest view of his beautiful plumage. The maneuver requires a good amount of stamina, much like acrobatics on a parallel bar. If the female is impressed, she will lean forward and peck him on the beak as a signal to mate. When finished, she flies away and he begins calling for another female.

The magnificent bird of paradise builds his own stage by creating a clearing, removing twigs and other debris from the forest floor with his beak, and stripping leaves from branches to allow sunlight in to shine on his iridescent feathers. When he is satisfied with his arena, he begins vocalizing with coos and whistles to attract female passersby. Once he has captured the interest of a female, he begins a series of flashy moves, bowing to raise his long

Bowerbirds are the master architects of the animal world. They build elaborate bowers to attract females for mating. Young bowers must learn from older males how to build with precision artistry. Males decorate these bachelor pads with colored glass, plastic, shells, and paint their walls with brightly colored berry juice. From an evolutionary perspective, bowerbirds have replaced elaborate ornamental feathers with these structures. The most plainly feathered bowerbirds build the most elaborate bowers.

On an expedition to Australia from 1838 to 1840, explorer John Gould discovered the bower of the spotted bowerbird. Gould made this painting of the structure, believing it had been built by aborigines as a playpen for their children.

tail feathers in the air and spread his yellow cape. He combines all of the common moves into his routine—bobbing, swaying, and dancing with the up-and-down step found in so many other species. He vocalizes with a harsh buzzing song.

The most ingenious lekking behavior is found in the bower-birds of Australia and New Guinea. The ornithologist E. Thomas Gilliard, who was fascinated with the bowerbird's attraction signals and displays, said that birds should be split into two broad categories—bowerbirds in one, and all other birds in the other. While many species of lekking birds clear their stage of leaves, grass or debris before getting on with the show, bowerbirds go further. From sticks and grass, these fellows build structures of varying complexity that are the animal world's Playboy mansions. The most creative and spectacular builder in the animal kingdom is the plain brown bowerbird. Somewhere along the bowerbird evolutionary highway, it stopped investing in bright colors and transferred its visual displays from body ornamentation to the bower, the structure it builds to attract mates. All the bowerbirds are skilled weavers. Their constructions could easily have provided the models for early hominid classes in mat weaving 101. Take a basic brown bowerbird design, turn it on its side and add a bottom, and you've got a basket.

Bowerbirds vocalize so that females can locate them, and they do some strutting on cleared stages, but the main attraction is their bower. No matter which species builds a bower, it is constructed so that the front and rear openings are oriented to the north and south. The male is ensuring that the sun will not blind the female when she gazes at his masterpiece.

Satin bowerbirds have a particular fondness for blue. They have brilliant blue irises and perhaps their distant ancestors had more colorful, bluer feathers than today. The male satin bowerbirds paint the inside walls of the bower with the juice of blueberries and weave into the walls a variety of man-made blue objects. They also roll out the blue carpet, so to speak, by lining their front walkways with bits of blue glass, flower petals, and blue plastic. The most colorful bowerbird, the Australian regent, builds the simplest bower and does not decorate it at all. The MacGregor's bowerbird builds what is called a maypole bower and creates a raised circular court around the structure. The design uses a central sapling around which the bird piles twigs. The male then hangs colorful ornaments from the twigs. The result strongly resembles a Christmas tree.

The most elaborate bower is the one built by the plain bowerbird. It starts out with a central maypole but resembles a hut by the time it is finished. The first Europeans to discover this type of bower were convinced that primitive humans built it. The brown bowerbird has elevated decorating to a true art, using anything he can find that is colorful and aesthetically beautiful, including butterfly wings, flowers, pink seashells, berries, and man-made objects.

Altogether, Australia is home to 8 species of bowerbirds and New Guinea to 18, all of which have different ways of attracting females to the bower and of entertaining them. The spotted bowerbird calls for a female and then stands off in nearby bushes to await his guest. When the female arrives she steps into the bower and inspects the craftsmanship. The spotted bowerbird paints his walls with the juice of red fruit mixed with saliva and prefers white objects, including bone and shells, to line his walkway. If the female is satisfied, she vocalizes for the male to approach. He approaches on his walkway, which is also his stage, and begins vocalizing and dancing. If all goes well with the courtship, she will consent to mating with him in the bower.

Lainy Day, of James Cook University in Australia, says the bower is one of the rare forms of communication in which the signals are expressed using construction. Some people would be loath to describe the decorative bower as art, since a bird has created it, but the parallel is inescapable. All bowerbird species have a penchant for shiny and colorful objects, which are used in the walls and walkways of their structures. Day's research shows that great bowerbirds have a particular affection for green, white, gray, and red. As decorators, these birds place their grays, whites, and greens at the ends of the bower and use reds on one side and around the perimeter. Apparently, the color scheme makes the male's gray and black feathers contrast nicely against the backdrop of the bower.

Scientists at the University of Maryland have found that females tend to be sensitive to aggressive vocalizations and dance moves of the male and that the male requires a good deal of experience to assess whether he is coming on too strong or not strong enough. This research team is famous for its mechanical female bowerbird and its studies of the male's reactions to the robot's responses to his courting abilities in the wild. The "fembot" is programmed to simulate the female's behavior during courtship. The Maryland group's research shows that courtship is reciprocal and dynamic between the male and female and can no longer be described simply as male performs, female judges, female chooses. A lot more is going on between the two sexes than scientists generally had assumed.

Normally a female satin bowerbird flies overhead, inspecting the overall construction of different bowers from her bird's-eye view. When she sees one of interest, she lands and steps inside to inspect it. The male then pops out from his hiding place onto the carefully constructed stage in front of the bower and begins his vocal and visual song-and-dance number. The male satin bowerbird's display is especially noisy and aggressive, but if he is too aggressive the female will become frightened and fly away. If he is not aggressive enough, she will become bored and fly away. Research on bowerbirds shows that the more experienced males receive the most female guests and the greatest number of consensual signals to mate. The female satin bowerbird signals her consent by crouching.

These intelligent birds progress through stages of learning to become successful lovers. The first step for young males is to learn how to build the bower. Inexperienced males must find an older mentor and, like young carpenters, undergo an apprenticeship. After they observe older males building their bowers, the young males attempt their own. Their first bowers are rather clumsy, lopsided efforts. Over a period of several years, their skills improve and they can construct their own elaborate bowers. Males frequently steal the most prized decorative objects from each other. When a male leaves his bower, he might return to find that another male has trashed it. There is some evidence to suggest that young males are even recruited by older males to engage in some of the vandalism and larceny, which would explain why older males tolerate the presence of future competitors at their building sites.

Gerald Borgia of the University of Maryland, who has studied bowerbirds for more than 20 years, has developed an intriguing theory regarding the evolution of intelligence. In nature, motivations for behavior are rather simple. An animal is either attempting to have sex, conquer or defend a territory, achieve status and dominance, or get dinner. Borgia suggests that the primary motivation for intelligence in bowerbirds is competing for and securing a mate. His research shows that the most intelligent birds are the best bower builders and the most sensitive to the responses of females during their displays. As a result, they are the ones who attract the largest number of mates. One successful male mated with 25 females over a two-month period, while nearby builders of lower-quality bowers landed 0 mates. Another expert bower builder mated with 9 females in one day.

Visual signals for attracting mates in primate species are much less conspicuous. Here, the lines between female choice and male choice become blurred. Species that engage in monogamous relationships and "one-male breeding systems" rarely display overt visual signals and tend to have more flexibility in the timing of sexual interactions, according to Sarah Hrdy, of the University of California at Davis, and Patricia Whitten, at Yale University School of

Medicine. Females in multimale systems like those of olive baboons and chimpanzees display swellings of the perineal region that are unambiguous visual signals. These species tend to mate around the time of ovulation. Females of other species usually have the upper hand when it comes to choosing mates. Barbara Smuts, of the University of Michigan, has proposed that females can choose directly, by responding to opportunities to mate, or indirectly, by influencing which males are permitted to join a group.

Females have a number of subtle visual signals at their disposal for initiating sex. Depending on the species, females will crouch, flick their tongues, pucker their lips, shudder with their heads, or display the clitoris. Sometimes females will be quite direct. One report from observations of brown capuchins described an estrous female who followed the dominant male around for several days. Throughout this "shadowing" period, she would frequently approach the male, grimace, make a distinctive vocalization, push him on the rump, and shake branches at him. When she was ready to copulate, she chased him until he stopped running, and then they mated.

Female Japanese macaques hop onto the backs of males and display mounting motions to signal their interest. Females of all primate species also exert control by simply refusing to copulate. The only evidence of forced copulations in nonhuman primates comes from observations of orangutans, in which solitary males sometimes form a coalition against a lone female. In all other cases, despite the persistence of some males, sex is consensual, according to Smuts. Females refuse by walking away or by screaming and becoming aggressive. In savanna baboons, females sometimes leave their groups and consort with males from other groups. In one instance, a female brought her new partner back to her group, which he was allowed to join. In gelada baboons, an outside male will sometimes try to take over a group. His success depends on whether the females are attracted to him. If they consent to copulate with the outsider, the current leader surrenders his dominance to the new fellow, who becomes the leader. Infanticide, in which a male kills young off-spring to induce females into estrus so that they will copulate with him, has been observed in various species of primates. Males are also known to become aggressive and violent with females. In chim-panzees, females will often mate with multiple males in a group to reduce their aggression toward her and her offspring.

A sexual invitation from a male chimpanzee might include branch shaking and a beckoning gesture toward a female. Generally speaking, dominant males have the mating rights in a group, but primates often show preferences for individuals that may run against the typical dominance hierarchies. Harold Gouzoules was observing a group of macaques in which a clandestine mating occurred. A female in the group had come into estrus and was being shadowed by the dominant male. Wherever she went he was right behind her to ensure that no other male would gain access. After a few days of this, the female had climbed a perch and was just hanging out, the dominant male at her side. Maintaining eye con-tact, she carefully withdrew and climbed down to take a drink at a small watering hole. The male watched her very closely, tensed, and was ready to leap down should any other male try to approach. The female all the while kept turning to maintain eye contact with the male. She returned to the perch, sat for a while, repeated the trip to the watering hole, and returned again. By this time she had either gained some trust with the dominant male or he let down his guard. On about the third time that she climbed down to take a drink, the male dropped his vigilant watch for a brief moment and the female quickly disappeared into the bushes, where another male was waiting. They were able to copulate before the dominant male threw a fit and chased them both. Gouzoules said he never saw any overt signaling between the female and the other suitor, but they may have had this plan in mind. In most species, however, male choice boils down to a fight among the males, with the winner gaining the mating rights.

Primates appear to be unusual in their display of ornaments. In about 24 species of nonhuman primates, females advertise sexual receptivity with swellings of the perineal region. These vary from slight swelling and pink coloration to the conspicuously swollen red areas of female chimpanzees and baboons. The swellings appear when females are ovulating, and are found in females of old-world apes and monkeys—those from Africa and Asia—but not in new-world species, found in Central and South America. These swellings of perineal skin are an overt signal involved in mating, but until recently it was not known whether they indicated a female's degree of fertility or her health. A study conducted in Africa by Leah Domb, of Harvard University, and Mark Pagel, of the University of Reading in England, is apparently the first to show that genetic fitness is the message being inferred by males.

During the Victorian era, out of embarrassment, people covered up females' swollen bottoms at zoos, but the bigger and redder the swellings, the more willing the males are to fight for the females. Observations of olive baboons at Gombe National Park in Tanzania revealed that females with larger swellings reached sexual maturity at earlier ages, gave birth to more offspring, and produced offspring with the best chances of survival. Males fought each other more often for the big-bottomed females than for females with less pronounced swellings. So the sexual swellings of female baboons appear to be an "honest" signal, like the peacock train, because they accurately reflect reproductive fitness in females.

Perineal swelling has disappeared in human females, although adult human females have permanently swollen breasts, a primate oddity unrelated to fertility. However, the breasts may swell slightly during ovulation. The absence of a visible signal of fertility provides human females with a distinct advantage by allowing them to select when they want to mate without regard to their reproductive cycle. The human female's ability to disguise outward visual signs of reproductive receptiveness and to mate whenever it suits her plays a powerful role in attracting human males to become monogamous partners who will provide paternal care and other resources.

Scientists have been curious whether ornamentation of other types comes into play in human male-female attractions. Bobbi S. Low, a scientist at the University of Michigan, writes in her book *Why Sex Matters: A Darwinian Look at Human Behavior* that ornamentation is abundant in both men and women, and that visual displays used for mate attraction cross most human cultures. According to Low, men use ornaments—such as fancy cars, expensive clothes, jewelry, and even cell phones—in 87 of 138 societies that were studied. A study recently conducted in India found that young, poor men buy inexpensive, fake cell phones to carry around as displays of status. Another more recent study found that American men display cell phones in bars as a mate attraction ornament and that they are even more likely to use their cell phone in the presence of other men as a status or power signal.

Low found that women do not display ornaments as much as men, but female ornamentation is prevalent in 49 of 138 societies that were surveyed. Cosmetics, however, are a universal means of ornamentation in women. I witnessed a renaissance for women in Kabul, Afghanistan, in the winter of 2001–2002. Under the Taliban, all women were forced to wear the burqa, a piece of clothing that covers the body from head to toe. Nail polish, beauty shops, and cosmetics were forbidden. Within a couple of weeks of the Taliban's defeat in Kabul, the beauty shops were reopened and cosmetics began arriving by the truckload from neighboring Pakistan.

After physical exercise and during sexual arousal, a woman's cheeks tend to be flushed, so rouge naturally follows as a visual signal intended to convey fitness and sexual desirability. During arousal a woman's lips also redden. Lipstick is therefore used as a visual signal across many cultures. David Givens, director of the Center for Nonverbal Studies in Seattle, points out that lips may be considered an ornament since they are turned out and reddened to draw attention.

One marked difference that Low found between the displays of men and women is how they advertise marital status. Low says that women in more than 70 percent of cultures advertise their marital status, while men in less than 10 percent of cultures do the same. Polygamy is the rule in 83 percent of human societies that have been studied. We share our polygamous nature with the majority of other mammals, including birds. Even the prairie vole, which was considered to be the most faithful of species, engages in discreet tomfoolery, according to DNA studies of the pups. Monogamy is a choice and a social contract between two people, not necessarily a natural state.

Givens suggests that humans have evolved more subtle nonverbal expressions to communicate attraction. To describe one of these, he quotes Shakespeare's Romeo: "See! how she leans her cheek upon her hand." Givens calls this the head-tilt-side, which can express friendliness, foster rapport, or show coyness or "cuteness" during courting or submissiveness during aggression. According to Givens, as primates we have more in common than we might have imagined with the animal kingdom when it comes to visual mate attraction signals.

"Bright colors, floral prints, bold lines, and geometric shapes attract the eye," Givens writes, "as do necklaces, bracelets, and watches that gleam in evening's dim lights. Designed to spot colorful fruits and berries from a distance, our primate eyes notice feminine necks decorated with strings of round, pigmented beads, which catch eyes and whet a desire to reach out and touch what seems to be 'edible.' Worn as corsages, flowers designed to lure pollinating insects attract the eye, as well, and tempt our nose with their sweet fragrance."

Parties, Givens says, are human leks: men claim miniterritories, marking the immediate area around them with personal possessions such as newspapers, cell phones, and car keys that are set in the "reach-space" beside their drink glasses, finger food, and napkins. His "artifact scatter" is a signal that says "I am here," and from fixed courting stations like those of the bowerbird, he and his colleagues sit and stand noticeably erect, puff out their chests, tell jokes, posture, and laugh loudly, as if to say, collectively, "We are here."

Meanwhile, women cruise the party space, moving from table to couch to kitchen, to the restroom and back, skirting and brushing past the stationary men. A woman may seem to ignore them, but she is reading their nonverbal reactions to her movements and gaze. She preens, sweeps her eyes from side to side across a man's line of sight, glances back and forth, and circulates, Givens says. Her restless to-and-fro states "I am here."

I nearly rest my case. But let us move on from the territorial lek of the party to examine how the rest of the animal world is handling the real estate market.

Eight

OUR HOUSE

AT LEAST A DOZEN mosquitoes have been buzzing around my face and ears for the past hour. Even though I am sweating in the damp jungle heat, I have a blanket drawn tightly over my body and pulled up to just below my nose. It's 4:30 a.m. and for a change I am sleeping, or trying to, in a real bed in the relative luxury of the Posada de la Selva in the Petén rain forest, in northeastern Guatemala. The main trail leading to the grand Mayan ruins of Tikal is at my doorstep. I arrived late yesterday afternoon with a close Aussie friend after a long, bumpy ride on a mostly dirt road from neighboring Belize.

My friend is here to see the ruins, but I have come on business—to meet the rulers of the dense jungle that surrounds this ancient Mayan city. The black howler monkey, the largest nonhuman primate of the New World, is king here, as it has been since long before the stones of Tikal were laid nearly 1,800 years ago. For hundreds of thousands of years, these rulers have ferociously announced their presence and claimed their territories just before dawn. Guatemala is in the neotropics, so the sun rises at about 6. Even though it won't be up for another hour and a half, with mosquitoes greedily taking their blood meals, it seems like a good time to get rolling. I need to be sitting at the top of the Temple of the Two-Headed Snake, also known as Temple IV, at least half an hour before the sun shows its face. A couple of lukewarm Fantas provide a sugar rush while I shower myself with bug repellant.

Temple IV faces east and rises 200 feet above the jungle floor, towering over the rain-forest canopy. The tallest structure at Tikal, it is *the* place to be for sunrise. The Mayan pyramid is a brisk 20-minute walk from the lodge, so my friend and I are out the door, backpacks stuffed with breakfast bars, fruit, water bottles, binoculars, and cameras, and sitting on the top step by 5:15. This vantage provides a breathtaking view of the ruins of Tikal, once a city of 100,000 people. Like the Mayan civilization that vanished a thousand years ago, the black howlers are in danger of vanishing as well.

For the moment, thanks to the dollars tourism brings to the Guatemalan government, the jungle surrounding Tikal is protected, leaving this population of monkeys with a sufficient share of real estate. Black howlers can be found throughout Central America in pockets where the rain forest habitats they require are still intact. Costa Rica has a healthy population of howlers; some howler territories also remain, to a lesser extent, in parts of western Belize. Southern Mexico, Central America, and northern regions of South America are home to six species of howler monkeys.

As we sit at the top of the temple, a dozen coatimundis—the raccoons of the rain forest—begin milling about near the steps as more humans gather for the sunrise. Opportunists, they know breakfast is coming. The coatis raise their noses, sniff, and cozy up a little closer to a family that has just arrived. The little cons will soon have these people feeding them their breakfast bananas. Animals, like young children, are keenly adept at finding an easy meal whenever possible.

The glowing face of my watch reads 5:35. In the distance, from the south, comes the first pronouncement of the kings: "R-O-O-A-A-A-R-R-R-R-R!" It sounds absolutely ferocious, easily mistaken for a giant cat. Another ruler announces its presence to the north: "R-O-O-A-A-A-R-R-R-R-R!" Seconds later another call from the southwest: "R-O-O-A-A-A-R-R-R-R-R!" The chorus has begun: "I proclaim this territory. I am a powerful defender. All who hear me, heed my warning!"

Delivered from perches as high as possible in the canopy, the territorial calls of black howler monkeys can travel through the dense jungle for over a mile without degrading. The monkeys' need to be heard by distant neighbors presents one of the greatest challenges of all territorial species because of the thick vegetation that interferes with sound. The black howlers do not have the luxury of the open terrain of elephants or the SOFAR channel of the Goliaths of the marine world to make their long-distance calls. Like those species, however, the howlers produce their signals at very low frequencies, but in shorter, powerful bursts of sound that they sustain for several seconds. The burst-pulse helps to push the sound through and across the vegetation. These morning rituals are not turf wars so much as security reminders from one dominant male to another to stay where he belongs. The monkeys make these calls before dawn every day of the year as long as they are alive. It is perhaps the most important job of their entire day. If one of these kings fails to make his announcement, the others will know something is wrong, that defenses are down and territory is ripe for takeover.

After the dominant males have begun howling, the females and their offspring often chime in. Each territorial group has at least one male, two to three females, and some offspring. The size of the groups varies with the amount of unbroken habitat available to them. Howlers need a lot of space, and when they have it the groups can be as large as 20 to 30 individuals, consisting mostly of extended families. Howler territories can range in size from about 3 acres to as much as 25 acres. The largest of these ferocious howlers is about the size of a two- or three-year-old chimpanzee. The adults of the species range in size from about 20 inches to 36 inches from head to rump. A unique physical feature of howler monkeys is their long, quite muscular prehensile tail, which they use like a third hand for grasping. Great apes dropped their prehensile tails long ago on the evolutionary trail, but unlike many primates, howlers do not have opposable thumbs, which explains why the tails are so important. When howlers travel they use their tails to maintain a grasp on one branch until they've secured the next one with their hands. Sometimes adults link their tails to form bridges between distant branches for youngsters to cross.

Looking at these smallish creatures, I find it hard to believe that they can make such a tremendous sound. But they have evolved a natural vocal instrument: an enlarged hyoid bone in the throat, and saggy, folded jowls that they fill with air like a bagpipe to achieve a powerful burst. This morning, the howlers keep up the territorial calls until about 9. Each group listens carefully to its neighbors and most assuredly recognizes the individual voices of its border rivals. That females and the young join the calls as a chorus provides information about group size and the relative strength of the home-land defense to the neighbors and any free-roaming unmated adult males. A howler monkey can easily tell the difference between an adult male, the females, and the young: male adults are bigger than the others, and thus their vocal instrument produces a lower fre-quency. Hearing multiple frequencies, or harmonics, allows listen-ers to determine that the roars are coming from a group, as well as how many members belong to it. If a family has several young, it may permit unrelated adult-sized males or more than one unre-lated adult male to join its chorus to increase security. Their com-bined voices will make potential invaders think twice before attempting a takeover. After the morning chorus, the groups settle down, roaring only when they encounter another group near a ter-ritorial boundary.

As soon as young males are mature, they are usually made to leave the natal group and make their own way in jungle life. Single males form bachelor groups, not unlike wild stallions, and they eventually challenge dominants for their territories and females.

The howlers are not very vocal with family members during the day, except for occasional soft grunts that might mean "Move over" or "Come here." One well-documented signal is the flicking of the tongue, which males and females use to solicit sex from each other. Black howlers socialize less than other primate species and spend a lot of time lounging on upper branches in the canopy. The leaves they eat do not provide them with high energy, so they conserve it by not moving around much. If humans approach below their territories, the family may move a bit higher and the dominant male will come out to investigate, but because of the thousands of tourists that come to explore the ruins, the monkeys have become accustomed to the paparazzi wanting to shoot their pictures. Let a single howler male approach, however, and the whole family will begin roaring and rattling the branches.

Across the Atlantic Ocean, another turf war is raging, in the forests of northern Europe. The speckled wood butterfly is a species that places an equally high value on its territory, which happens to be sunspots created by shafts of sunlight that filter through the trees and spill onto the forest floor.

When a sunspot appears, a male speckled wood butterfly will immediately claim it and defend it against other males that try to enter. Possession of a sunspot, however fleeting, greatly enhances the male's chance of attracting the attention of a female speckled wood butterfly. The sunlight reflects off the male's wings and allows it to stand out in the shaded forest like an actor under the spotlight on a darkened stage. The sunspot is the speckled wood butterfly's lek. In addition to making him visible, the sunspot offers the valuable resource of warmth to both the male owner and the female. The solar energy helps these butterflies rev up their metabolism after being in the colder shady climates of the forest.

When a male comes across an unoccupied sunspot, he often

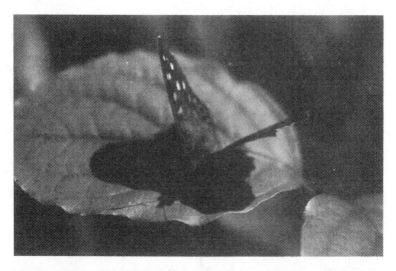

The speckled wood butterfly considers sunbeams peeking through Northern European forests to be prime mating territory. It will occupy and defend sunspots against intruders.

claims it for as long as it lasts. If another male happens to enter the sunbeam, they engage in a brief aerial battle in which they spiral around each other upward into the light. When the battle is over, the intruder simply flies away. The owner always appears to win these contests. The noted ethologist N. B. Davies conducted experiments to see what would happen if he removed the original owner from its territory for a few minutes and allowed the intruder to take over. When the displaced owner returned to the sunspot to reclaim its turf, a second contest ensued, with both butterflies spiraling again, but in these experiments the displaced owner always departed. For this butterfly species, it is finders keepers on the turf. Animal communication scientists call this behavior the bourgeois strategy, meaning ownership is respected and owners never lose challenges by intruders. Rodents also abide by the bourgeois strategy, using pungent scent markings to declare territories, which other rodents respect.

A group of skeptics in Sweden conducted their own territorial sunspot experiments with the same butterfly species in a similar forest and found the opposite result: the displaced owner engaged in a much longer spiral flight when it returned to the sunspot and won it back from the intruder. Why the different results? It turned out that the studies were conducted at a different time of the year, when the amount of sunlight streaming into the forest changed considerably. Davies conducted his study in the late spring, when sunspots on the forest floor were more plentiful. The Swedish group conducted its study in early spring, when the sun is low in the sky, the nights are longer, and sunspots are less plentiful. To explain the difference in behavior, Davies reasoned that supply-and-demand economics raised and lowered the value of sunspots to owners. With a greater availability of sunlight, the value to the butterfly was low, so the displaced owner could save his energy and move on to a vacant sunspot. But with less available sunlight, the value was high, so the displaced owner fought harder to win it back.

Sunspots provide direct value to owners by increasing the butterfly's body temperature. The longer a displaced owner was prevented from returning, the colder it became and the less likely it was to win back the turf. The longer the resident had possession, the more it warmed up and was better able to win an energetic aerial contest. The moral of the butterfly studies is that possession of territory conveys special advantages to owners that increase their odds of winning turf battles. The owner of a territory in most cases will fight harder than an invader because the value of the real estate to the owner is usually higher.

Territorial behavior is nowhere more evident than in songbirds. A classic example of songbird territoriality is offered by a subspecies of song sparrow—a small, scrubby-colored bird. The males, which have a truly impressive repertoire, are highly territorial, and springtime is a particularly noisy season as they stake out their claims. After wintering in Mexico, the males must acquire about a quarter of an acre where their mates can build a well-concealed nest on the ground or in a bush, and where the family can obtain all the bugs, berries, wild fruits, and seeds they will need for the season.

As one sparrow arrives, he surveys the area and settles on a territory that suits his needs. He will begin to call "tweet tweet doodle tweet." It is still early in the season and the territory is by no means secure. Another male settles into a bush nearby and hears the territorial claim. The contest is on. The second male repeats the call: "tweet tweet doodle tweet." The first caller realizes that he must compete and this literally ruffles his feathers. The two will begin a song-matching bout that both have trained for all their lives. The first sparrow calls back to communicate what Eugene Morton translates as: "My territory. I am great." The two repeat this back and forth, deliberately annoying each other and trying to judge how tough the other guy really is. They assess each other's voices and song-matching skill and determine their respective locations. They may also be able to determine without seeing each other something about their relative sizes and social status. When one male decides that he can win a physical battle, he takes wing and charges his competitor. They will fight to the death if territories are scarce and there is no alternative.

As the season progresses and the territories have been established, neighbors take up a new tactic with each other. Individuals possess about 6 to 12 songs and know 10 to 15 variations of each one. After the boundaries are established, a male sings the song of his neighbor, but usually only when the other male is not singing, This is a less threatening way of maintaining territory and means something like, "I'm so great I can sing your song, so don't forget who I am." To the uninitiated human standing in the backyard with the morning coffee, the songs sound quite sweet. But these guys are all talking trash.

Bradbury and Vehrencamp have worked out a nice summary of the advantages of owning real estate that tip the balance of conflicts in the animal owner's favor:

1. Owners have proven their ability to acquire property and have a big investment in protecting their offspring and nesting sites on all-purpose territories (territories where the animals nest, breed, and forage).

2. The holders of prime breeding spots attract the most partners for mating. *Prime* for an animal means there is safety from predators, good defensibility, and nearby food.

3. If the territory is the place where an animal normally forages, the animal already knows the location of the ripe fruit or best prey items, and it knows all the great hiding places to stash food. The territory also gives the owner the advantage of being able to ambush an invader.

4. Territory owners have already established their boundaries with their neighbors, which provide stability and make it easier for residents to know when strangers are in the area. An alert owner might overhear his neighbor growling, "Get off my property." If the owner is a bird, it might see a neighbor fly off to a distant boundary to challenge an interloper. The owner can raise its security level by increasing patrols and refreshing scent marks.

Real estate, whether it is an expanse of rain forest or a sunspot on the forest floor, is important to just about every living creature on the planet. It is a good bet that members of most of the species on earth have had some dispute in the past 24 hours over turf. All of us, from the ants beneath our feet to the monkeys in the trees to the birds flying overhead, are obsessed with real estate. Even in the newspaper business, we use territorial analogies and squabble over our territory every day. Reporters are usually assigned a specific beat, such as politics, or the Pentagon, or in my case, science, to which we refer as our turf. In my section, turfs are divided into subgroups, so that one reporter covers infectious diseases, another covers psychology, and another women's health. When a reporter from one turf steals a story from another reporter, we call it "bigfooting"—as in leaving footprints like the legendary yeti. If a reporter bigfoots a colleague, he is isolated socially and is not invited to lunch, and surrounding reporters raise their security levels by not discussing story ideas that an interloper might steal. Disrespect for another person's turf is a tremendous source of agitation and seething conflict in the workplace, no matter what the occupation.

As primates, we can take our turf battles to violent extremes. Before coming to Tikal, I was hoping to get into a more remote region of the jungle, but Guatemala has dangerous primates that carry AK-47 machine guns and possess a nasty reputation for raping, robbing, and murdering other human primates who stumble into their territory. Of all species, people are unsurpassed in their proclivity for violence. Much murder and mayhem are due to extreme territorial behavior. *Homo sapiens* behaves as if nature had bundled the fierce territoriality of songbirds, fire ants, and chimpanzees into one volatile package. Perhaps we should rename ourselves *Homo cidalmaniacus.*

Besides the human-against-human war, Guatemala has been the scene of a terrible territorial war between humans and the rain forest and its inhabitants, including the descendants of the Mayans. Ranchers backed by large U.S. companies grabbed the land and cleared the forests for grazing cows for fast-food hamburgers. These companies claim they don't do this kind of thing anymore, but there was a time not long ago during which the indigenous populations were slaughtered along with the forests and the wild animals. This is a tired old story that is still being played out in nearly every country on earth. Humans need more pasture, more space for villages and gardens, more strip malls, more trees for lumber, more, more, more. We don't pay nature a dime for all this real estate either, or repair and restore it after we've finished with it, even though every inch of ground that we seize is home to numerous species of plants, animals, and insects. Oops, there goes another rain forest. Oops, there goes another watering hole for elephants. There goes the neighborhood of a hundred pairs of songbirds.

Humans are not good at sharing. Perhaps some people hate any evidence of their evolutionary roots in the animal kingdom. Whatever the reason, the human species is winning an ultimately fruitless, self-destructive territorial war against nature and simultaneously remodeling the surface of the entire planet.

Other animals are not above killing each other over territory, but they are much less inclined to escalate to that extreme. Generally, they follow rules of engagement that have been worked out over millions of years. All species have developed clear signals to identify their boundaries. Chemicals play a strong, if not pungent, role in marking boundaries. Lacking the olfactory sensitivity of animals, humans erect fences and post no-trespassing signs. Chemicals are signposts that convey ownership without requiring the owner's presence. Dogs are masters of scent marking and surveillance. I do not know any dog than can walk 20 feet without sniffing the ground to check out which canine rivals have been on its turf. The sexy pheromones employed for mate attraction are the wrong variety for marking territories because they dissipate too quickly, so mammals manufacture and disperse low-volatility compounds by rubbing specialized glands against surfaces or spraying them on the ground, bushes, or rocks.

The perimeter of a territory is heavily marked so that neighbors will know where the boundaries are located. But that is not enough for a truly effective homeland defense. It is always possible that another animal could innocently pass between perimeter signs and enter the territory. So for added security, the animal, whether a deer, rabbit, or fox, always marks the trails of its property and any internal boundaries surrounding nesting sites. All scent markings are refreshed regularly. Many mammals also emphasize their territorial scent marks with visual signals of strategically positioned feces that stand out against the background, typically at the intersections of paths, on large flat rocks, and around trees.

Established neighbors generally respect these marks because everyone has to tend to the business of life—attracting mates, raising families, and getting groceries—which requires time and energy. Nevertheless, an animal must make time in its day to patrol its borders and keep a keen eye and ear on the neighbors. Security is about being seen, heard, and smelled by the neighbors, the animal equivalent of leaving the lights and the television on. My mother's dog, Taco—fat, old, and tiny as he is—maintains a fierce defense of the backyard, even though it has a high privacy fence. He lives indoors, but every few hours or so he waddles to the back door and vocalizes his familiar "arf!"—his signal to be let out. Unfailingly, he bolts out the door as if chasing an intruder, barking with a serious, no-nonsense tone. He patrols the perimeter, refreshes his marks, and then stands in the middle of the yard and barks a few more times. "I am Taco and this is my yard." Then he comes to the door and arfs politely to be let in.

Animals' colors, spots, and stripes are primarily a means for attracting members of the opposite sex, but their second role is for neighbor recognition. Because they are a little different on every animal, the markings help identify individuals. Two animals on a border patrol can easily recognize each other and not waste energy gearing up for a fight. If a stranger appears, the owner of a territory must take action immediately and issue a threat signal. On territories defended by monogamous pairs, males threaten other males, and females threaten other females. The owner needs to close the distance between himself and the intruder to get a better look and size up the situation. The markings that help females size up potential mates also allow potential rivals to compare their sizes. Cats and canids will walk side by side to determine which one is bigger. Even if the intruder is bigger than the owner, the owner has the home advantage and may be able to recruit family members or social group members.

In a potential border conflict, animals assume their aggressive postures and use their low-frequency growls. Every species has its own quirky way of saying, "Just try me!" Ring-tailed lemurs typically engage in stink fights when a conflict escalates. Each rubs its long striped tail across scent glands in its wrists and then faces off with a glare, aiming its pungent-smelling tail at the opponent. When a domestic cat detects an interloper in its territory, it arches its back, raises its fur, and issues the guttural cat growl. The confrontation is won by the cat that maintains its growl the longest and does the best job of varying the growl's intensity and frequency.

Wild horses do not defend a permanent territory, but in some societies they aggressively defend their harems and whatever food or water resources they are using. Animals that range, such as horses and elephants, try to assert dominance when two groups arrive at a watering hole or special grazing area. Their size or their previous experience with each other in a conflict might determine which group has dominance. The subordinate group will back away and wait for the others to leave. Older dominant stallions usually control harems of several females and their offspring. As soon as male offspring are about a year to two years old, the father chases them away before they start getting ideas about mounting the females. Single males form bachelor groups within which they form alliances with one or two other stallions. The bachelor groups chase each other and bite and kick in play fights to build their skills and strength for the day when they may have to fight a dominant male to take over his harem. When two wild stallions confront each other, they come face-to-face to assess each other's size, then suddenly jerk their heads up and unleash terrifying screams intended to tell each other that fighting is futile: "I am powerful. Walk away while you have the chance."

Ears are important for sending visual signals in many mammals. Horses, like dogs and wolves, lay back their ears to signal aggression. If an animal possesses weapons, such as claws, antlers, or strong hind legs that can kick the bejesus out of an opponent, it displays them to its opponent to say, "Are you talking to me?" Stallions begin a conflict by running and taking turns chasing each other to show their endurance. If one is not intimidated by the other's strength and does not back down, they begin biting, often tearing large chunks of flesh from each other's shoulders. In the final stages, they back up to each other to kick in an effort to sever each other's leg tendons. Ginger Katherines, a documentary filmmaker and an expert on wild horses, captured on film a long, violent fight between two stallions in which the defender's rear leg tendons were cut, making him unable to give chase as the challenger made off with his harem. The defender was permanently disabled.

Primates in an aggressive encounter begin by staring with the evil eye. Chimpanzee mothers, when needing to reprimand offspring, have been observed to take hold of a daughter's face and forcing her to look when being scolded. As my mother would say, "You look at me when I'm talking to you!"

If an animal species claims a permanent territory, it has invested its life in that property. Everything, including its mate and offspring, is at stake when the territory is threatened. The more serious territorial conflicts arise when single males mature and begin searching for a place to settle or conquer. Lone males are the greatest threats in mammal and bird species that establish permanent or breeding territories. Some single males may be inclined to kill the resident male's offspring and start a new kingdom with his females. Infanticide sometimes occurs in howler monkey troops, as well as among elephants, horses, lions, chimpanzees, humans, and other primates. The practice is found more often in polygynous species in which the offspring live at home with their mothers for a relatively long time.

A female in a monogamous relationship must stay alert for unmated younger females that may attempt to displace her and take her mate. Resident females become quite agitated when an unmated female comes near their borders. Duetting and chorusing with their mates during territorial announcements signal to other females to stay away.

Chimpanzees on a border patrol behave like the gun-toting primates in Guatemala and many other countries. Adult male chimpanzees can weigh up to 110 pounds and are twice as strong as a human. Chimpanzees at the Gombe Stream National Preserve in Tanzania, where Jane Goodall conducted her studies, frequently patrol the borders around their community's range. If they hear strangers in the distance, the males will immediately investigate. Chimpanzees also are known to kill males from other communities to expand their borders. Goodall reported that the Gombe chimps on one occasion killed every male of a neighboring group. Several years ago, the dominant male at Gombe grabbed the baby of a camp worker and smashed its head, killing it. The scientists there decided not to kill the chimpanzee because it was acting on its natural, albeit violent, instincts.

Like the black howler monkeys, Australian magpies are among the species that defend a permanent, all-purpose territory. An all-purpose territory is a like a ranch where nesting, mating, and feeding all take place. The territories of Australian magpies usually include multiple stands of trees with wide-open spaces in between. They live in cooperative groups that can include extended families and unrelated birds that came for dinner one day and never left.

It is not clear to scientists whether male and female magpies form monogamous bonds as a rule. Some territories have been found to contain only a single nest that is tended by the entire group, while others contain numerous nests occupied by pairs. Either way, females usually take care of the young while males take responsibility for foraging and feeding the females on the nest. Males also share some parental duties.

No matter what social strategy a magpie group settles on, all members of the group participate in territorial defense. Neighbors constantly skirmish and try to expand their borders, and flocks of

migrants land and try to take over the ranch. Can you imagine settling into a neighborhood, working out the boundaries of your yard
with your neighbors, and then coming home from work one day to
find 20 people camped in the yard, eating all the food from the
refrigerator? When that happens to a group of magpies, their highly
territorial neighbors will not lift a feather to help out. For this reason, magpies rarely travel beyond their borders. If they did, they
would quickly lose the farm.

Australian magpies have developed a sophisticated means of territorial defense, called caroling. Magpie songs have both warbles
and carols, with warbles consisting of short, soft, fluid notes and
carols using extremely loud, long sounds. The warble is sung first
as a solo and may be a way of saying, "I see potential trouble. Let's
gather round and sing a carol." The warble is followed immediately
by the carol, sung either as a duet or by every bird in the group.
Males and females have sex-specific versions of the carols, and each
individual has its own unique notes thrown into its sex-specific
songs, but magpies don't develop predictable, stereotyped song
repertoires like songbirds in North America and Europe. A stereotyped song is one that is always sung the same way. Magpies are
more like rappers. Intelligent and capable of learning new songs
and mastering new sounds throughout their lives, they like to
improvise and are skilled mimics, sometimes incorporating sounds
from other species into their songs, which soon spread to other
members of the group. Communal groups of magpies share song
syllables that they have learned from each other, but individuals
maintain a large degree of unique song syllables as personal identi-
fiers. The shared song syllables of carols establish a group identity
that is employed for territorial defense, but the carols also reveal the
identities of individual members, the sex of each member, the
number of members, and the identity of the group. The softer warbles in the broad repertoire of magpie vocalizations are used mostly
when a group is preening, hanging out, building nests, and feeding

the hatchlings. Ravens and crows also use softer vocalizations that are highly variable and individual when they are socializing as a group. The caws that people imitate are territorial defense calls.

During turf battles with neighbors, magpies engage in countersinging matches similar to the song matches of songbirds in temperate zones. The magpies' superior learning abilities allow them to learn the songs of their neighbors and sing them back to the neighbors as a way of intimidating them: "I can sing your song better than you." Most songbirds from temperate zones stop learning new songs when they reach adulthood, but Australian magpies, most crows, jays, and ravens in temperate zones continue learning throughout their lives. One reason for this might be that songbirds in North America and Europe establish single-nest territories and, during the nesting season, do not form large social groups, while crows, ravens, jays, and magpies are highly social and frequently mingle with nonthreatening newcomers. When new members are allowed to join a social group, they first need to learn the group's social password—the group chorus. This is also true in tropical species of parrots.

The ornithologist Eleanor Brown kept a pet magpie named Zoe for a number of years. They typically greeted each other with a familiar whistle that Zoe had learned as a youngster. When Zoe was three years old, Brown began making a new whistle during greetings. After two weeks, Zoe learned to imitate the whistle perfectly and began using it whenever Brown entered Zoe's turf—her aviary. Brown said that Zoe had adopted five adult humans with whom she had regular contact. They all used the common whistle during greetings, and Zoe always allowed them into her aviary without a fuss. But whenever a stranger entered the aviary she would become fiercely territorial and attempt to engage in countersinging with the human stranger. Since Zoe's repertoire consisted primarily of human whistles, she would use those whistles in her territorial matches with strangers.

Some species' version of territory is the mating site, such as the lek of a sage grouse, or a nesting site, such as the spot to which albatrosses return on Midway every year. These turfs are defended as vigorously as any permanent territory. Fights are not usually violent, but they do sometimes break out, with males biting each other. The prime real estate is along the atoll's western beaches in the native *naupaka* plants, which provide thick foliage roughly four feet high to block the sun. There is also plenty of room between the trunks of the hearty shrubs for maneuvering along the ground. More important, the ocean breezes easily pass through the *naupaka* near the ground to cool chicks during the hot days of spring and summer. But any place on the atoll that is not paved becomes some couple's turf as the birds begin to nest. Cracks in the abandoned runways on Eastern Island are also gaining favor as nesting sites among young adults, like subdivisions in the burbs of a bustling city. Males that have not yet bonded will often stake out a nesting territory in these less desirable, unclaimed regions. If a young turk or some dim fellow who is confused about his turf has taken an owner's spot, the owner will squawk and the new fellow will leave—owner's rights.

Fish can be surprisingly territorial. The grouper, which can live 20 or 30 years or more, is a highly territorial fish that never strays too far from home. One grouper became something of an old friend to officials of the National Oceanic and Atmospheric Administration (NOAA), who are responsible for protecting the National Marine Sanctuary Program, which includes the sunken wreck of the Civil War ironclad USS *Monitor.* The wreck is located off the coast of North Carolina, about a two-hour boat ride from Cape Hatteras, in an area known as the Graveyard of the Atlantic. It sits on the bottom at a depth of about 230 feet, where the grouper has claimed a shelf created by the *Monitor*'s hull.

When I was covering a story about an operation headed by NOAA and the U.S. Navy to raise the *Monitor*'s giant gun turret and conserve it for a museum, I had the opportunity to travel down to the *Monitor* in a three-person submersible, called the *Johnson-*

Sea-Link II, operated by the Harbor Branch Oceanographic Institution. The sub was lowered into the ocean by a crane from a ship about half a mile from the site. Once underwater in the sub, we followed a compass heading to the wreck site along the bottom. The water is pretty murky most of the time around the *Monitor* because of a strong current, but on this afternoon it was surprisingly clear. I was sitting in a bubble with the pilot at the front of the sub, and as we slowly approached the *Monitor* the grouper swam out quickly from beneath the stern to challenge us for invading its turf. The pilot was focused on making sure he did not bump into the wreck or one of the navy divers who was working at securing a giant claw to the gun turret, but as soon as I saw the grouper I forgot about the wreck. Groupers can grow quite large, up to four feet or more. Once you meet one eye to eye, you may not want to order it for dinner again. This grouper saw us approach and must have figured that it was in for some kind of fight with the sub, which might have looked like a one-eyed sea monster. Even though the sub was considerably larger, the fish held its ground in front of its turf and charged the sub several times as a warning to move away. If the grouper was grunting, as it and many other fish species are known to do, it was not audible through the submersible's acrylic dome, but the fish's body language quite clearly signaled aggression. As the submersible passed into the grouper's territorial space, the fish became increasingly aggressive, and it kept its eyes on us the entire time we were at the site. Since the sub did not charge back but floated by without an incident, the grouper may have regarded its defense of the homeland as a victory. Some fish will battle for a nesting site by locking jaws and pushing each other in a contest that can last longer than a scuba tank of air can keep a diver down watching the performance.

Inadvertently, I engaged in a song-matching contest and won a turf war with a house sparrow that was just settling inside the air-conditioning unit of my bedroom. A couple of Christmases ago, one of my sisters gave me something called a Royal DM70ex organizer, a calculator with an alarm clock and a storage system for phone numbers. This unit gave me a false sense of technical expertise, though I dared not whip it out at a bar as an ornament to try to impress women. During one of my disciplined periods of working on this book, I had set my Royal DM70ex organizer alarm for 5:30 a.m. so that I could write for several hours before leaving for work. At this time of day, a thin line of deep red forms on the horizon across the Potomac River from my bedroom window. As my alarm began chirping "beep beep . . . beep beep . . . beep beep," the sparrow seemed to be responding with a "tweet . . . tweet . . . tweet" for the full minute that the alarm sounded. At first I thought it was a coincidence. The second morning I realized that the bird was competing with the alarm in a song-matching battle for the air conditioner. On the third morning, very agitated growling was coming from inside the air conditioner. I should have turned the alarm off right then, but for some reason that I will always regret, I didn't. The duel began as it had before, but now the sparrow fluttered and banged around inside the air conditioner as if it were trying to punch its way into my bedroom to attack the organizer. Unable to break through the face of the air-conditioning unit and kill the enemy, the sparrow flew off. For the first spring in seven years no bird settled in the unit. Coincidence perhaps. But I believe I had witnessed a turf war won by a piece of plastic. And yes, I felt sad that spring when the soft begging calls of hungry little hatchlings did not drift into my ears each dawn.

Now the time has come to move once again to the ocean, to explore the communications of some of the most intelligent species on earth—dolphins and their killer whale cousins. Although birds can learn and mimic many sounds to communicate, these marine mammals are surpassed only by humans in their ability to learn. If we want to discover the origins of human language, we should carefully examine these mammals and a phenomenon known as vocal learning.

Nine

ALL IN THE FAMILY

MORE THAN 13,000 years ago, the first people arrived at Santa Cruz Island, to the west of California, the largest of the eight Channel Islands, from somewhere across the Pacific. The Chumash civilization settled the islands and, for thousands of years, crisscrossed these waters in long wooden plank canoes known as tomols. According to Chumash legends, their people were created on Santa Cruz Island, known then as Limuw, where the heavens touch the sea. When the population grew too large for Limuw, their guardian, the goddess Hutash, transported the people across a rainbow bridge to the mainland, where they flourished again for thousands of years. During the migration over the rainbow, many of the people gazed downward and became disoriented by the height and the deep fog below. Losing their balance, they tumbled into the sea. Despairing over the loss of so many people, Hutash transformed the lost souls into dolphins. The Chumash regard dolphins as their brothers and sisters and still call them the people of the sea.

The sun is disappearing behind Santa Cruz Island as the *McArthur*, a 175-foot research vessel, cruises slowly northwest away from Star Crazy Island, towing a sophisticated sonar system that is mapping the seabed. The deep rose of the California sunset blends with the dark slate of the flat calm Pacific Ocean, creating shimmering coral pastels on the surface. Most of the crew is below deck eating dinner. I am standing at the bow listening to the ocean wash gently against the hull when my ears prickle at a faint, high-pitched squeal. Ed Cassano, manager of the Channel Islands National Marine Sanctuary, shouts, "Dolphins!" Cassano calls me to the starboard side just in time to see several brothers and sisters of the Chumash streaking toward the *McArthur* like silver torpedoes. Another half dozen sleek bodies glide playfully alongside the ship, then dart beneath the bow. More dolphins appear, breaking the surface just in time to dive over the bow wave. Dolphin voices begin to fill the air as hundreds of the graceful animals emerge from the east

Common dolphins often travel in herds of hundreds or thousands of members. They are found throughout warmer waters of the world's oceans. Dolphins may be one of the few animals that have developed individual "names," known as signature whistles. They acquire their individual signature whistles shortly after birth and appear to keep them for life. Family members and closely bonded groups use signatures to communicate, similar to the way people call each other's names.

swimming toward Santa Cruz Island. I ask Cassano where the dolphins are going. He shrugs, but suggests with a grin that they might be planning to stalk the fishing trawlers that come out at night, lights blazing, to skim the boundaries of the sanctuary where the marine life is plentiful and supposed to be protected.

The din of dolphin voices increases as more appear. We become mesmerized by their sheer numbers, their calls, and their purposefulness. They seem to know where they are going and why. As they pass the ship in no particular hurry, infants swim playfully ahead of their mothers, who swim with sisters, cousins, and other adult females. Small alliances of two or three males swim and dive side by side in perfect synchrony. Cassano estimates that more than 1,000 dolphins are traveling together toward their mysterious destination. The Chumash people must have understood their dolphin kin quite well after living as neighbors for so many thousands of years. It is easy to imagine the laughter of their children as dolphins escorted the 30-foot tomols on their journeys between the islands. No doubt the rainbow legend was repeated by the adults many times on those occasions.

For as long as humans have had contact with dolphins, we have understood that we share a unique bond. The Chumash possessed a deep respect for the dolphin character and regarded the animals as equals. The Chumash culture was overrun by Spanish conquerors more than 400 years ago, but a contemporary group of people locked under the spell of dolphins is striving to gain a new understanding of their behavior, intelligence, and ability to communicate.

In this final chapter, I will explore some of what has been learned about the communicative abilities of our brothers and sisters of the sea and suggest where we might look to find even greater similarities between the innate natural language shared by animals and humans, and what we call human language. The complexity of dolphins' and other marine mammals' talk has been attributed to vocal learning, which essentially is the ability of an animal to modify its vocalizations based on its experience with other animals. Marine mammals are masters of vocal learning. Through their voices we can begin to trace the sounds of our own.

The story of marine mammal research opens in 1953 with Kenneth S. Norris, generally regarded as the grandfather of whale and dolphin research, who had just been hired as a young founding curator of the oceanarium Marineland of the Pacific. Norris's fascination with dolphins led him to confirm that dolphins possess sonar similar to bats' and can "see" by means of sound waves. Their big brains are wired for sound in the same way that human brains are wired for vision. Dolphins can form three-dimensional sound images by bouncing sound waves off objects and analyzing the echoes, and their resolution is so finely tuned that they can locate a silver dollar underwater at a distance of nearly 250 feet. Captive dolphins can distinguish between a BB and a kernel of corn at 50 feet. They are so adept at sensing texture that they can distinguish between sheets of copper and aluminum.

At about the same time Norris was working on echolocation, John C. Lilly was studying neurophysiology and the human brain as a medical doctor at the National Institutes of Health. After inventing the isolation tank and the electroencephalograph machine to measure brain waves noninvasively, Lilly became intrigued with the idea of floating 24 hours a day and was soon exploring the ocean world of dolphins. He left NIH in the late 1950s and began studying dolphin whistles and sonar, convinced that dolphins were speaking a complex language just like humans. In 1961, Lilly launched dolphinmania with his popular book *Man and Dolphin,* which began, "Eventually, it may be possible for humans to speak with another species."

Lilly never proved that dolphins possess a language, much as he tried, but his genius (and perhaps a bit of madness) created new tools and methods for studying dolphin signals. Both Norris and Lilly made tremendous contributions to the field of dolphin communication. Norris's more conservative scientific approach appealed broadly to other scientists and attracted men and women to the field of marine mammal research the way Jane Goodall attracted a new generation into primate studies. But Lilly's bold style did more to capture the public's imagination. His work at developing a common language for dolphins and humans became the basis of the movie *Day of the Dolphin.* Later on, his experiments with hallucinogenic drugs in the isolation tank became the fodder for the movie *Altered States,* an accomplishment that did not aid his credibility with more conservative scientists. Nonetheless, Lilly never stopped trying to communicate directly with his dolphins. The following is from a 1983 interview with Lilly published in *Omni* magazine:

> I just did a very primitive experiment—a Saturday afternoon–type experiment—at Marine World. I was floating in an isolation tank and had an underwater loudspeaker close to my head and an air microphone just above me. Both were connected through an amplifier to the dolphin tank so that they could hear me and I could hear them. I started playing with sound—whistling and clicking and making other noises that dolphins like. Suddenly I felt as if a lightning bolt had hit me on the head. We have all this on tape, and it's just incredible. It was a dolphin whistle that went *sssshhheeeeeoooooo* in a falling frequency from about nine thousand to three thousand hertz in my hearing range. It started at the top of my head, expanding as the frequency dropped, and showing me the inside of my skull, and went right down through my body. The dolphin gave me a three-dimensional feeling of the inside of my skull, describing my body by a single sound! I want to know what the dolphin experiences. I want to go back and repeat the experiment in stereo, instead of with a single loudspeaker. Since I'm not equipped like a dolphin,

I've got to use an isolation tank, electronics, and all this non-sense to pretend I'm a dolphin.

The question of whether dolphins are speaking a language sim-ilar to our own has not been resolved, but dolphins and their larger cousins, the killer whales, are highly intelligent. Whistles, burst-pulses, and click trains are the dolphin's and the killer whale's repertoire of vocal signals. Whistles are very high frequency sounds, most of which are above the range of human hearing. Burst-pulsed sounds are the squawks, squeals, snaps, cracks, bleats, barks, moans, and groans for which dolphins are so famous. Click trains are the sounds dolphins make when using their echolocation system. Often, when a dolphin is communicating with another dolphin, it makes click trains and whistles at the same time. One theory about this dual vocalizing is that the whistles are conveying information while the click trains are sensing the other animal.

Typically, click trains are used for navigation and hunting. When dolphins go out for dinner they emit sonar clicks at a rate of about ten per second—the staccato sound is the click train. That rate increases as a dolphin moves closer to an object. When dolphins approach each other using echolocation or are investigating an object at close range, the clicks increase to 400 per second and sound like a steady tone to the human ear. Processing those kinds of signals and converting them into 3-D images require serious brain-power. Dolphins can also vary the frequency and energy levels of their sonar to adjust for background noise.

Scientists have published hundreds of papers describing in re-markable detail the structures of these types of vocalizations—their bandwidths, their harmonics, and their energy levels—but relatively few papers can be found that describe their meaning. We may never know whether dolphins speak a language, but many of the qualities that people consider to be human, at least since the time of Descartes, certainly exist in dolphins and other marine mammals.

Most research since Norris's first experiments has been conducted on bottle-nosed dolphins. All species, however, converse socially through vocalizations, percussive slaps with their flukes against the water, and body language. Dolphins emit all of the sounds they make, except their sonar clicks, through the blowhole. They produce their ultrasonic clicks with a separate organ called the melon, which is located directly beneath their rounded foreheads. The sound waves appear to pass from the melon through a nose plug or maybe through the mouth. The dolphin sonar receiver is its hollow lower jawbone, which detects vibrations and transmits the sound to the inner ear.

High-frequency ultrasonic waves can be used as a weapon. A captive killer whale once released a burst of sound above the surface and knocked a trainer on his back. It was probably not an accident. Among their own kind, dolphins and killer whales are more polite and usually turn their sonar off when another dolphin or whale passes in front of them. Peter Tyack of Woods Hole has suggested that the primary reason dolphins swim side by side when foraging may be to avoid interference with their signals and to prevent their sound waves from zapping each other. When conversing, dolphins appear to take turns with their whistles and click trains.

Like other cetaceans, including humpback whales and killer whales, dolphins slap their tail flukes against the water to create percussive sounds that are communicative. No one is certain what they mean, but they may be an aggressive signal. Because dolphins do not have olfactory bulbs and nerves, which terrestrial mammals use for the sense of smell, chemical communication among them is thought to be quite limited except for their sense of taste. Scientists have some evidence that dolphins do communicate via pheromones, which they detect through taste at very close range.

Dolphins have excellent vision both in and out of the water, even though they rely more heavily on sound. Their eyes have evolved special pigments that are shifted toward the blue end of the color spectrum to increase their perception underwater. Scuba divers find that much of the color of visible light is dissipated underwater and can compensate for the loss of color by wearing dive masks with pink glass. To human eyes, which are evolved for detecting light in a terrestrial environment, everything under about 25 feet looks rather pale and blue. Dolphins use their vision in part for recognizing kin and social acquaintances. Individuals have distinctly different coloration and markings, which serve as visual signals in the same way that patterns in a land animal's fur or feathers do. Teeth marks from fights show up as light streaks on a dolphin's skin, suggesting that the bearer is a veteran of many battles; they might serve as a badge of dominance in some species.

The most dramatic visual displays found among dolphins and whales are bubble streams, which they emit while vocalizing underwater. As they make sounds, they push air through the blowhole. The bubble streams may be coincidental or produced purposely so that others in the water can see which dolphin is talking. Coincidental or not, dolphins can identify speakers by their bubble streams. Humpback whales that have attracted a female will sometimes release a stream of bubbles up to 90 feet long as camouflage to prevent other males from spotting her. Bubble "nets" and bubble "clouds" are used for rounding up prey. The bubble clouds startle fish and exploit a natural reaction to danger, which is to group together. Schooling fish, cows, sheep, and ungulates all have the same grouping instinct when they sense danger. Normally, grouping reduces the odds of being attacked by a predator. But dolphins and whales exploit the innate behavior of prey to their distinct advantage. Humpback whales and killer whales produce bubble nets to create barriers through which fish cannot swim. Killer whales do this together to trap hundreds of fish and then take turns swimming through the concentration to eat. Humpback whales use bubble clouds to round up fish like cowboys herding cattle. Adults may teach this hunting behavior to younger animals as a part of cetacean culture.

At the New England Aquarium in Boston, scuba divers enter the dolphin tanks periodically to perform maintenance. One of the dolphins earned a reputation over the years for blowing bubbles in a pattern that perfectly imitated the bubbles escaping from the scuba diver's regulator when the diver was working. Since dolphins naturally make bubbles when they are vocalizing underwater, it is possible that this dolphin was imitating the scuba diver's bubbles as a way of saying hello. Mimicry is commonly used by higher social species as a means of addressing each other or showing group membership. In the wild, sometimes a dolphin will approach a scuba diver in a friendly manner. If the diver does a couple of somersaults, the dolphin might imitate the diver's moves, taking them as a signal to play or interact. Humans use imitation both vocally and visually to communicate when they do not understand each other's languages. Why couldn't dolphins and humans?

Dolphins have been shown to have the ability to learn artificial languages such as those taught to great apes. They possess mental tool kits, as primatologist Mark Hauser of Harvard University would say, that provide them with a basic understanding of grammar and syntax. Lilly created computer sounds that dolphins could use to communicate via touch pads or keyboards. He also discovered quite early that dolphins can mimic human speech.

The dolphin whistle has captured perhaps the most attention in studies of communication. Dolphins in captivity readily learn and mimic each other's signature whistles, and they imitate the pattern of the dog whistles used by human trainers. We might assume that dolphins interpret the whistles made by trainers as something akin to the trainer's name. To a dolphin, my whistle would be the equivalent of saying, "I'm Tim." The dolphin's imitation of my whistle might be translated as, "Yo, Tim."

When the dolphins were passing by the *McArthur*, the air was filled with the whistles, used as contact calls, between females and their offspring and between the males in the small alliances. Most social species on land use contact calls as well, although in many species they appear to be innate. Dolphin whistles are learned. As a rule, contact calls are short and only loud enough to be heard by the caller's intended receiver. If two people were walking together in the rain forest and were out of visual contact, every now and then one of them might yell, "Hey! Are you still there?" And the other might respond with a simple "Yo!" But if there were 40 or 50 humans trekking along on a big expedition, yelling "hey" is not specific enough to get each other's attention, especially if other people are shouting too. In this situation, the two individuals might use each other's names. This appears to be what dolphins do when whistling.

When animals become separated and cannot find each other, they typically make a signal known as an isolation call. These vary in their form depending on the species, but overall, an isolation call carries with it a degree of distress. According to Jack Bradbury, the common response of most terrestrial mammals is for the listener to move immediately toward the signaler. But scientists discovered as early as 1965 that when captive dolphin mothers are separated from their calves, or when members of a social group are separated, they increase their use of a type of whistle unique to individual dolphins. The dolphins continue whistling until they are reunited. These calls are known as signature whistles and resemble the isolation calls of terrestrial mammals and birds in their sound structure. No two whistles sound alike, though whistles among family members share some features. This level of individuality is possible only because dolphins possess a talent for mimicry and learning new vocalizations. Bottle-nosed dolphins learn their signature whistles when they are very young and probably keep them throughout their lifetimes. Scientists have documented the whistles for as long they have been able to track a single dolphin, about 12 to 15 years.

Only in the past few years have scientists learned that wild dolphins intentionally match each other's whistles to communicate. When an infant wants to say "Hey, Mom!" it calls its mother's signature whistle. The mother responds with her calf's whistle. Similarly, males in alliances appear to call one another's whistles when they want to find out where the others are located. Dolphins are the only species other than humans believed to use names to identify individuals.

"We knew bottle-nosed dolphins learn sounds, and there have been occasional reports that dolphins copy each other's calls," says Vincent Janik, a behavioral scientist at the University of St. Andrews in Scotland, who has led most of the recent research on signature whistles. "We did not know whether this was done in the wild or what the function might be. But if you can copy another animal's signal, then you can address that animal."

Dolphins live in a complex, large society. Related dolphins form small pods, and unrelated pods gather together and swim in large herds that can include hundreds of animals—like the one at the Channel Islands National Marine Sanctuary. During a typical day, dolphins separate from their companions or relatives as they forage and may spend hours away from each other. Pods often spread out over a mile or more. Janik says that while foraging, mothers match the signature whistles of their offspring to locate them or ensure they are nearby. Males might do the same when they've been foraging separately and are ready to move back together.

In the late 1990s, Janik conducted a groundbreaking study of signature whistles, off the coast of Scotland in a channel frequented by bottle-nosed dolphins. Eavesdropping on groups of dolphins is no easy task. If dolphins are aware that they're being watched, they alter their normal behavior. To study dolphin vocalizations, researchers usually deploy a hydrophone from a boat or from the shore and record the sounds, but this does not help to distinguish between individuals or to determine whether they are communicating with each other. To overcome this limitation, Janik cleverly deployed three hydrophones in a triangular formation and recorded small groups of dolphins for nearly six hours. By applying a mathe-

matical formula that included the distance between different whistles, the maximum swimming speed of dolphins, and the speed of
sound underwater, Janik was able to sort out when one dolphin was
communicating and matching another dolphin's signature whistle.
Using statistical analysis, he determined whether matching signature whistles occurred more often than by chance and found that
dolphins not only actively match signature whistles, they also begin
vocalizing afterward, when they may be sharing additional information. What that additional information might be is still not
known, but Janik is continuing his studies at St. Andrews to seek
the answers.

"Such individually distinctive, learned signals for individual
recognition so far have only been found in humans," Janik says.
"The occurrence of such referential labels is thought to be a significant step in the evolution of human language . . . Captive dolphins
can use learned sounds to label objects and can be trained to report
on whether objects are present or absent by using these labels."

Janik is among the small group of researchers who are aggressively exploring the meaning and behavioral contexts of signature
whistles and the role that vocal learning plays in their development.
Dolphin infants learn their signature whistles within the first few
months of life though imitation and experience. Male dolphins
learn signature whistles that are quite similar to those of their
mothers; female babies learn whistles that are distinct from Mom's.
Young male dolphins stay with their mothers for up to six years
before moving off on their own, and they do not sexually mature
until they are about 20 years old. After leaving their mothers, males
form alliances that can last for a lifetime and that for all practical
purposes are the same as pair-bonds between members of the
opposite sex. DNA studies show that brothers—actually half brothers—often form alliances. If males learn signature whistles similar
to their mother's, then it is possible that male siblings have signature whistles that are similar, which might enable little brother to

find big brother when he leaves home. This is not proven, by any means, but scientists say it would be interesting to pursue as a research question.

Janik and Peter J. B. Slater say that evidence for vocal learning in mammals is found only in whales, dolphins, harbor seals, humans, and greater horseshoe bats. The scientists suggest that vocal learning arises under the conditions of sexual selection, defense of resources, and the need for individual recognition in large complex societies. The complex society of dolphins makes signature whistles a very handy tool with which small groups can locate each other.

Dolphins are extremely tactile creatures and engage in a great deal of sexual behavior during their social gatherings, which is not widely reported by scientists, probably because it is a bit X-rated. Because their sexual, sensual nature plays such a large part in dolphin social life and communication, it warrants discussion. In the wild, dolphins express themselves tactilely with wanton abandon. Scientists refer to these tactile expressions as sociosexual behavior. Dolphins mount each other quite regularly for reasons other than mating. Most of the mounting appears to occur between members of the same sex. Young males attempt to mount their mothers within weeks of birth, and males mount each other as juveniles and adults. Females, usually young or pregnant ones, also mount each other and rub genitalia. Scientists say that mounting appears to "mediate social relationships" by helping the dolphins form bonds, but it is also a way of expressing dominance and aggression. The meaning of mounting behavior varies with the situation, but it occurs between all ages and sexes of dolphins.

When dolphins "fusion" after being separated for some period, they greet by rubbing their dorsal fins, which may be the equivalent of a dolphin handshake or hug. Greetings among most species of mammals and primates involve intimate touching of some variety. Male chimpanzees often greet each other by touching and holding each other's testicles, which implies a certain amount of trust. Since dolphins are so tactile, it seems reasonable that they would rub each other in areas that provide pleasurable sensations as a way of bonding. Humans would do something similar if we had not developed spoken and sign language as a means of stroking each other's egos with compliments and pleasantries, but we do hug and kiss each other like chimpanzees and other primates. Clasping hands is also a primate gesture. The human tradition of waving to a person at a distance is thought to be based on a signal that says, "Look, no weapons. I come in peace." Dogs and wolves nuzzle each other's open mouths when they greet. They also mount each other to assert dominance. Gentle physical contact is the antithesis of attacking and it conveys the message of friendliness unambiguously.

Dolphin skin is quite sensitive, especially around the eyes, jaw, blowhole, flukes, and genital areas, which appear to be as sensitive as a person's lips, fingertips, and genitals. During social encounters, dolphins often stroke each other with their flippers and sometimes swim together with one dolphin resting a flipper on his or her companion. All creatures need tactile stimulation. Studies in humans show that massages lower heart rate, respiration, and blood pressure. Cats purr when petted, dogs get droopy eyes and open their mouths to signal contentment, humans sigh and moan during a massage, and dolphins, killer whales, and large baleen whales vocalize softly during their social-sexual encounters.

Male dolphins commonly caress and stroke each other with their fins, flukes, beaks, and heads while rolling, chasing, pushing, and leaping, according to Bruce Begemihl, the author of *Biological Exuberance: Animal Homosexual and Natural Diversity.* This behavior can last for hours and is often accompanied by erect penises. The female spinner dolphin has been observed riding another female's dorsal fin by inserting it into her genital slit. They then swim together in this position. Female bottle-nosed dolphins actively masturbate each other by taking turns rubbing each other's clitorides. Both males and females have genital slits. Male bottle-nosed dolphins have been observed engaging in something called beak-genital propulsion, during which one male inserts his beak into the genital slit of another male, then propels both of them along with gentle thrusts of his head. Male bottle-nosed dolphins also chase male Atlantic spotted dolphins and sometimes engage in mounting behavior. These encounters may result in temporary alliances between the members of the two species.

Researchers reported an interesting use of the click train between an adult male spotted dolphin and a juvenile male of the same species. In this instance the older male produced a vocalization called the genital buzz by directing a stream of low-pitched, rapid buzzing clicks toward the genital area of the male calf. Also an aspect of heterosexual courtship in this species, this sound may serve as a form of acoustic "foreplay," actually stimulating the genitals of the recipient by means of the strongly pulsed sound waves, Begemihl says. Dolphins, it would appear, invented the vibrator.

The high level of sociosexual behavior in dolphins is thought to indicate a high level of cognition. The only other species that engage in such creativity—beyond the purpose of bearing offspring—are primates, including humans, bonobos, and, to a lesser extent, chimpanzees.

When dolphins do mate, males can become very aggressive in the search for females. Initially a male or group of males signals to a female with a loud, low-frequency pop. This appears to be used exclusively as a signal to females. Normally the female responds to the aggressive popping call by immediately swimming to the male. If she does not respond, the males in the alliance may charge her and even attack. The male alliance will also swim aggressively to isolate one female from her social group for mating. Scientists have observed females issuing vocal warnings to each other when aggressive males are spotted swimming toward an unsuspecting female. Sometimes these warnings give the female just enough notice to escape. At the surface, males sometimes engage in a series of less aggressive visual displays as a team in the presence of a female. One of these is known as the rooster strut, during which the male arches his head above the surface and bobs it up and down. Another courting maneuver is the butterfly display, in which two males swim figure eights on either side of the female in synchrony. Scientists speculate that these displays have something to do with impressing the female with their vitality.

With this type of aggressive behavior as the norm in dolphins, it is easy to see why swimming with them in the wild can be risky. Dolphins are powerful animals and can inflict serious internal injuries on each other. They attack by ramming their heads into another animal's torso and biting. A person who is swimming near a group of dolphins and hears popping sounds needs to get out of there or one of the dolphins may try to attack him or her.

Regardless of the species, aggressive encounters usually start with simple displays and only escalate when neither animal backs down. Such displays in terrestrial mammals involve body language, facial expressions, and vocalizations. Dolphins are no exception. In 1994, Hauser proposed a graded system, called predictive signaling, for aggressive signaling in dolphins. In a typical aggressive encounter, two dolphins begin with a face-off. Because the dolphin body is so sleek, threat postures are limited to expressions using the head, mouth, and flukes and arching of the back. Arching the head and flukes downward is a threat posture in bottle-nosed dolphins. One dolphin initiates the aggression by assuming the downward threatening posture and emitting rapid burst-pulsed sounds with his mouth closed. The sounds convey aggression. If the receiver does not immediately swim away, his response to the initial aggression is to remain facing his foe and emit his own burst-pulse sound. These low-frequency sounds follow the same motivation-structural rules that serve aggressive vocalizations in land animals. The next stage is to open the mouth, a signal akin to a dog's baring its teeth, and emit the pulsed sounds more forcefully. Dolphins can easily regulate the amount of energy they put into a burst-pulse. If neither of the animals gives up, the next phase is to shake the head up and down with the mouth open while continuing to emit the burst-pulses. By this point, the situation is quite serious. The last warning before an attack is the jaw clap, which is made by opening and closing the jaws very rapidly.

No creature in the ocean is more feared or perhaps more intelligent than the killer whale, *Orcinus orca*. Killer whales are also, without a doubt, the most socially stable species on the planet and the most widely dispersed mammals, next to humans. Just as we occupy all of the world's landmasses, orcas occupy all of the oceans, and, like us, they are at the top of the food chain. Killer whales are brilliant vocal learners. They develop sophisticated vocalizations and speak with dialects learned from their mothers and grandmothers. The dialects, like killer whale societies, are very stable. A killer whale's dialect reveals its membership in a specific familial pod. According to research conducted by Hal Whitehead of Dalhousie University in Nova Scotia, these animals' culture is handed down over successive generations. Some groups actively teach their young to hunt, and they cooperate with each other with amazing coordination to track and kill their prey.

The story of the orca is actually a tale of two whales—the fish eaters and the mammal eaters. In the frigid sea off of the coast of British Columbia, some of the most active killer whale research has been conducted for more than 30 years. The killer whales that live in the waters adjacent to Alaska, British Columbia, and Washington State are perhaps the best studied in the world, although the killer whales that grab seals and penguins off beaches and ice floes in the Crozet Archipelago in the French Antarctic and off the coast of Argentina have also received attention.

Paul Spong initiated the first study of wild killer whales at the Vancouver Aquarium in 1970, after observing and working with them in captivity for three years. Of his experience, Spong wrote, "I concluded that *Orcinus orca* is an incredibly powerful and capable creature, exquisitely self-controlled and aware of the world around it, a being possessed of a zest for life and a healthy sense of humour and, moreover, a remarkable fondness for and interest in humans."

Spong discovered that orcas display a special curiosity about music and will sometimes approach boats playing music simply to listen. John Ford, of the Vancouver Aquarium, and Fisheries and Oceans Canada, followed Spong's path and has become one of the leading scientists involved in killer whale research, studying the two populations in the northern Pacific. Ford calls one group the residents and the other group the transients. The residents, which have about 300 members, are a stable, highly vocal group that spends the summer and autumn primarily in the waters near British Columbia. They feed exclusively on fish and especially favor chinook salmon for their dinner. Residents have the most stable social structure known for any mammal, even more stable than those of humans—at least in Western society. No one leaves home. The males are referred to as mama's boys because they stick to their mothers' sides all their lives. Considering females can live to be as old as ninety and males can live into their early sixties, that is a significant amount of stability. These family groups have dialects that are distinct from all other groups. Killer whales will assess an animal's dialect when considering it as a potential mate and will rarely, if ever, breed with animals that share the same dialect.

Transients, on the other hand, are mobile, stealthy hunters that travel between Northern California and southeastern Alaska. Ford estimates that about 250 transients visit the coastal waters near Vancouver. They live in groups of three to five and feed exclusively on other mammals. The groups appear to consist of the mother, her oldest male offspring, and a younger son or daughter or two. Younger males leave just before reaching adulthood, and females leave as soon as they mate to form their own family pod. Related pods appear to affiliate, and sometimes dozens of transient pods can be found socializing noisily at the surface. Pods of residents also congregate to socialize, but residents and transients rarely, if ever, mix socially. Their dialects are very different, and they give each other a wide berth when their paths cross, though sometimes they do get into fights for reasons that are not entirely clear. One theory is that the residents simply dislike having the transients invade their ocean real estate.

Mammal-eating killer whales prefer harbor seals as a main course, but they eat sea lions, porpoises, dolphins, and some species of large whales, even blue whales. They have been photographed with hammerhead sharks in their mouths. Recently, scientists discovered a group of killer whales off the coast of New Zealand eating stingrays and flipping their bodies into the air like Frisbees. When the opportunity presents itself, killer whales have been known to grab land mammals that get a little too close to shore.

Scientists are actively studying orca vocalizations, but not a lot of detail has been published. We do know that, like dolphins, their closest relatives, they make three general types of vocalizations: whistles, pulsed calls, and sonar clicks. Scientists do not have any evidence that orcas produce signature whistles similar to those of bottle-nosed dolphins. Tyack has said this might be due to the differences between killer whale and dolphin societies. Killer whale families are larger and more closely knit, and they spend most of their time together. Their dialects are sufficient to allow them to identify each other as kin. Ford's studies suggest that individual pods produce on average about a dozen distinct vocalizations that they share and use repeatedly in different contexts. The dialect of a pod, which remains stable over time, is handed down from generation to generation. Being astute vocal learners, killer whales from one pod can imitate the vocalizations of another pod, perhaps as a way of contacting each other, just as dolphins use signature whistles to contact individual dolphins.

Residents are by far the most vocal of the killer whales. They can afford to be as noisy as they want because their fish prey does not appear to be sensitive to their calls. While little is known about the specifics of killer whale signals, they do seem to be conversing when socializing. Ford says that vocalizations produced by one member of a pod elicits vocalizations from other members. When residents are foraging they split into smaller subpods that spread out over several miles. With hundreds of killer whales dispersed in the water and all of them vocalizing, distinctive dialects serve as badges of group membership. According to Tyack, sperm whales are the only other cetaceans that have been observed using group-specific vocal repertoires of signals.

Echolocation clicks, which are used primarily for detecting prey, may play a role in communication, but their function or meaning is not known. Pulsed calls are a different story, and Ford breaks these down into two categories—discrete calls and variable calls. The discrete calls are used most often as contact calls and isolation calls when traveling and foraging. When killer whales are swimming they often exchange the same discrete call to maintain the integrity of the pod.

Each of three pods of the southern resident killer whales that live in Puget Sound—the J, K, and L pods—has its own distinct dialect, which the calves acquire through mimicry. But the J, K, and L pods share some calls, suggesting that they had a common matriarch at some time in their distant past. None of these pods shares any features of their vocalizations with northern resident killer whale pods that live off the coast of British Columbia. Ford says that new dialects arise when pods eventually split into smaller units. As in humans, dialects become increasingly differentiated over time after the pods have separated.

Residents produce discrete calls at a high rate when meeting with another pod. No one knows exactly how these arrangements are made, but scientists suspect the pods make their initial contact over some distance—the loud calls of killer whales can be heard up to about 3 miles away. Since most species of mammals use loud calls to advertise for a mate and defend territory, these calls of killer whales probably also serve as mating calls and may serve a territorial function by helping pods avoid each other while foraging and thus avoid conflicts over resources.

Scientists have observed meetings of pods on at least 16 occasions near British Columbia. Typically, when two pods meet, the members of one family line up side by side and face the members of the other family. After a little more vocalizing they touch noses and then intermingle. Killer whales mate at all times of the year, with spring and fall the peak seasons for births.

The variable calls of killer whales, for which scientists have not been able to determine a pattern or isolate "discrete," repeated calls, are more of a mystery. Ford says killer whales use variable vocalizations most often when in close contact and socializing, as do the crows and ravens described by Bernd Heinrich. The variable calls of crows, ravens, killer whales, and other vocal learning species are of such a complexity that they might be akin to language. A lot of scientists suspect this is true but are reluctant to say so publicly without any evidence.

The complex whistles of killer whales appear to convey emotional content that follows the motivation-structural rules. Used for close-range communication, the whistles are actually tonal sounds that can be produced at different frequencies, amplitudes, and with different levels of emphasis. Killer whale whistles are similar to dolphin whistles, but they do not seem to have the same function.

The mammal-eating transient orcas are stealthy hunters, remaining extremely quiet when foraging. Their prey take extreme countermeasures to avoid contact whenever they have the slightest hint that transients have entered the neighborhood. Captive sea lions display absolute terror and panic to playbacks of transient killer whale vocalizations. They do not habituate to the playback even though they are in a pool and no killer whales are in sight, taking less than a few seconds to leap out of the pool and onto the pool deck. Scientists observe the same reactions in the wild when they play recordings of transient killer whale vocalizations using underwater speakers. Harbor seals, sea lions, and dolphins all display strong evasive behaviors. Seals hide from transients in kelp forests, which help deflect the hunters' sonar. Sometimes seals dive and attempt to hide in a crevice on the ocean floor, which is usually a serious mistake. Transients hunt as a team and coordinate their efforts. If one of them is able to trap a seal in a crevice, the team members take turns guarding the seal while the others return to the surface to breathe. Their tactic is simply to wait until the seal runs out of air and is forced by its survival instinct to make a mad dash for the surface. This type of team effort is made possible by learning, intelligence, and a means of communicating each other's intentions.

When hunting without echolocation, transient killer whales must rely primarily on their hearing to find prey, which is why they silently cruise close to the shoreline, where they know the odds of finding dinner are the greatest—seals spend a lot of their time on the beach or in the shallow waters near shore. A fair amount of evidence exists to indicate that mammal-eating killer whales teach their hunting techniques to their young. It is generally believed that animal teachers are rare and that learning among the young is accomplished through the observation of adults. But orcas are clearly an exception. Transient mothers, unlike human mothers, encourage their calves to play with their food. This orients the calves to the mammals that should be included on future menus. Often, mothers will wound the prey and encourage their youngsters to make the kill. One puzzle is how the young offspring of transients learn to stay quiet when foraging for seals along the coastline. Calves, like children, enjoy making noise. Scientists suspect that mothers somehow communicate this message to their calves.

The clearest evidence for teaching comes from the group of mammal-eating killer whales that hunt off the coasts of Argentina and near the French Antarctic in a manner never observed among the transients of the northern Pacific, probably because of the lack of steep-sloped beaches in the north. The southern orcas beach themselves, grab a seal, and then return to the water. A complex tactic, it is performed exclusively by female killer whales. Females are about nine feet shorter on average than males, so it is probably safer and easier for them to make these difficult maneuvers. Scientists have observed adult killer whales giving stranding lessons to youngsters in the absence of prey and showing them how to wiggle back into the water. When the youngsters struggle and are unable to get off the beach, the adults will nudge the students back into the safety of the ocean. Scientists have also recorded instances when killer whales attacked seals in the presence of youngsters and handed them the prey while stranded.

Apparently, not all females practice stranding to catch seals. Scientists observed one calf with a mother that did not hunt in this manner. This calf would seek the company of an unrelated female that did hunt and join her on her rounds. This female willingly taught the calf how to perform the stranding. When the lessons were over the calf returned to its mother's side and stayed with her at all times except when hunting school was in session.

Learning to recognize the sound of the dinner bell is a good thing if you happen to be the main course. Vocal learning plays a critical role in the wild for harbor seals near British Columbia, which must do their best to avoid contact with transients if they are to survive. The seals have learned to avoid the transients, but they never flee from the residents, which do not eat them. They know the difference because they can distinguish between the whales' vocal calls. Harbor seals learn the difference between the whales' calls by experience, a sophisticated form of learning in an animal.

To get to the bottom of how seals distinguish between residents and transients, scientists played a series of digital tape recordings to seals using underwater loudspeakers. The scientists first played sounds of transients and counted the number of seals that took off. They then played resident vocalizations and the seals stayed put. When they played a third set of calls, from harmless fish-eating killer whales from a different region, unfamiliar to the seals, the seals reacted as they would to their predators' calls. This indicates that harbor seals learn by experience, as humans do. Seals must have started out with an ingrained fear of all killer whales but learned to overcome that fear through experience with fish-eating killer whales that never attacked them. Distinguishing between the residents and transients saves energy for the seals, who do not have to panic every time they hear a killer whale.

Harbor seals normally live in small groups, hanging out near the coastlines in the colder regions of the Pacific and Atlantic oceans. Their learning ability provides them with a distinct survival advantage. The same ability, combined with a relatively rare talent for mimicking sounds, transformed a harbor seal into a Boston celebrity. The story of Hoover, the talking seal, reveals important clues about the development of language and how close species might be to someday conversing with one another.

Hoover was orphaned by his mother, who quite likely had been taken by a killer whale. At just a couple of weeks, the baby seal would have died without his mother to feed him and later teach him how to hunt fish and communicate. Hoover was rescued by a new set of human parents, George and Alice Swallow of Cundy's Harbor, Maine. George put Hoover in the bathtub and started him on a diet of ground mackerel, which Hoover did not seem to like much in the beginning. When the little pup outgrew the bathtub, George erected a small tent outside the kitchen door where Hoover could sleep. During the day, Hoover swam in the Swallows' pond. Whenever Hoover wanted company, which was often, he would thump on the kitchen door to be let into the house. George would greet him with a "Hello they-ah" in his thick Maine accent. George and Alice talked to Hoover and allowed him follow them around while they did chores. George began taking Hoover with him on errands into Cundy's Harbor in his car. Hoover developed a liking for sticking his head out of the window like a dog. It did not take long for Hoover to become a celebrity in little Cundy's Harbor, especially with the neighboring children.

It was the children who first discovered that Hoover could talk. The seal had been living with the Swallows at this point for about eight weeks. Like songbirds and human infants, which learn to mimic vocalizations during a critical early period of development, Hoover had been listening to George and Alice. A little after Hoover had turned two months old, a group of children that had been playing with him at a pond came rushing up to the house, shouting to George that Hoover had spoken, just like a human. George said at the time that he did not believe the kids, but not long afterward he met Hoover along the path to the pond, and Hoover greeted him with George's own "Hello they-ah." George and Alice were dumbfounded. At the time, no one had any idea that a mammal other than a dolphin could mimic human speech. Hoover had paid close enough attention to his parents to use his phrases in a context-appropriate manner. His repertoire included "Hello they-ah," "Get oveh heah," and "Whaddaya doin'?" Hoover also produced a delightful laugh.

By about four months of age, Hoover's appetite was exceeding the Swallow's grocery budget—George had named the seal after the vacuum cleaner because of the creature's voracious appetite. With much reluctance, George loaded Hoover into the car and drove him down to Boston to the New England Aquarium. The staff was pleased to have Hoover join them. After turning Hoover over and assuring himself that this young family member would be well cared for, George said, "By the way, he talks." The scientists at the aquarium nodded patronizingly at the old man and promised to take good care of Hoover. The scientists, of course, did not believe George any more than George had initially believed the children. But when Hoover started becoming attracted to the female harbor seals living at the aquarium, he began vocalizing as seals do to attract a mate. Male seals normally sing, but Hoover had never heard adult male seals sing, so he attempted to seduce the females with the vocalizations he had learned as a pup: "Get oveh heah" and "Whaddaya doin'?"

The staff scientists were surprised but immediately began to study Hoover's vocal abilities. Seals have a voice apparatus similar to a human's, and they have a big thick tongue, so they possess the technical equipment for speech. What was surprising was that seals also possess vocal learning ability. Hoover's legend grew in Boston, and he became the aquarium's star attraction. His fame was comparable to that of the Kennedys, according to some Bostonians. Aquarium president Jerry Schubel said that not only did Hoover mimic speech, he appeared to use his small vocabulary at appropriate times. Hoover would speak to visitors when they came to visit with his "Whaddaya doin'?" call, and, of course, he always sounded just like George Sparrow. In 1985, at the relatively old age of fourteen, Hoover developed a viral infection in his brain and died. But during his tenure at the aquarium, Hoover mated, despite his unusual calls, and had several offspring, including Cinder, Beanie, Joey, Spark, Amelia, and Trumpet. At least one of his grandchildren is following Hoover's tradition of mimicking human speech.

Mimicry is one of nature's true wonders. Australia is home to some of the most amazing mimics on the planet. Lyrebirds, butcher-birds, bowerbirds, and spangled drongos are just a few that have added human sounds to their repertoires to impress members of the opposite sex. While poor Hoover simply did not know how a seal was supposed to sing, these species perfectly mimic chain saws, car horns, other animals, and more recently, the rings of cellular telephones to display their prowess. A satin bowerbird mimicked a cell phone so well that several bird-watchers on a nature walk began reaching into their pockets for their phones.

Mimicry plays a vital role in the complex social lives of tropical birds. The orange-fronted conure, a parrot that lives in the forests of Costa Rica, produces a "chee" sound that appears to be specific to an individual. Parrots imitate the chee calls of others in nearby groups that they would like to join. After playback studies introduced an unfamiliar chee call to a group of parrots, the parrots began imitating the call after hearing it four or five times. Normally after parrots have flocked, they will change to another group call that signals the birds to fly off to a different location, but the parrots in the playback study became agitated when the loudspeaker was unable to switch calls and get with the group's program.

In species that form stable, year-round social groups, such as marine mammals, Australian magpies, budgerigars, ravens, crows, and parrots, vocal learning has evolved to coordinate their daily social calendars and help them maintain affiliations, network with potential new members of a group, and unite groups to keep the neighbors from encroaching onto their turf.

The big puzzle is why our closest primate relatives do not have the same vocal ability. Most scientists say there is little evidence to suggest that chimpanzees or bonobos are vocal learners, and efforts to teach speech to chimpanzees since the 1920s have failed. According to primatologist Charles T. Snowden, of the Department of Psychology at the University of Wisconsin, chimpanzees lack the anatomical features to pronounce the vowels sounds i, a, and u and cannot produce the phonemes g and k. Reconstructions of the vocal tracts of Neanderthals suggest the same. Marine mammals and birds evolved vocal learning independently through the pressures of complex societies. Yet chimpanzees and bonobos both live in societies as complex as any of these vocally gifted species.

Perhaps a lack of research has simply allowed vocal learning in these primates to remain undiscovered. Chimpanzees may have been dismissed as vocal learners prematurely. Richard Wrangham and his colleagues at Harvard University in Cambridge and Hofstra University in Hempstead, N.Y., conducted a study of "pant hoots" among male chimpanzees housed in two captive environments: one at Lion Country Safari and the other at North Carolina Zoological Park. The two groups, which had diverse genetic backgrounds, possessed distinctly different patterns of calls, but within the groups the calls were similar. The suggestion that chimpanzees may be able to learn and mimic new vocalizations came when a new male was introduced to the group at Lion Country Safari. The pant hoot is the vocalization that is probably most familiar to people. It is a loud call that begins softly with several "hoo, hoo, hoo, haa, haa, haa" kinds of sounds, then builds in volume to a scream, known as the climax, and finally lets down. These calls can be heard for over a mile in the wild and often produce a chorus. Chimpanzees use the pant hoot in a variety of contexts, sometimes to announce a discovery or to make contact after being separated and other times by dominant members of a troop to simply reassert authority. People who study chimpanzees in the wild begin to recognize individuals by the sound of the pant hoot.

The new male brought to Lion Country Safari was known to have a pant hoot that included his own addition of a raspberry sound, also known as a Bronx cheer. After a relatively short period of time, at least five other males in the group added the Bronx cheer to their pant hoots. This appears to be a clear form of mimicry not previously observed in any captive studies of chimpanzee vocalizations. Wrangham and his team said "the results suggest that the calls in each group converged in structure as a result of vocal learning . . . Taken together our data contribute to mounting evidence that social experience influences chimpanzee behavior." A single study does not provide proof of vocal learning in chimpanzees, but it does suggest the possibility and certainly indicates the need for more studies like this one.

One of the other reasons scientists may have had difficulty finding examples of vocal learning in chimpanzees is that they tend to
be more expressive with gestures than with vocalizations. Despite
their famous pant hoots and screams, chimpanzees are not nearly as
vocal as marine mammals or humans. They shake tree branches,
drum loudly on hollow logs, make faces, and wave their arms and
hands in ways that most likely have more meaning than we are
capable of understanding. As talkers, we pay more attention to
sounds than we do gestures. But some scientists suggest that before
humans developed spoken language, we had developed a sophisticated means of communicating with our hands. Mimicry can be
accomplished as easily through gestures as it can be with vocalizations. Early hominids could have taught stone tool-making and
skills as fine as threading a needle through mime and without uttering a word. None of the technological advances of our human
ancestors required speech. Some scientists believe that studies of
chimpanzee gestural communication might reveal a form of mimicry as sophisticated as that observed in the vocalizations of marine
mammals.

At least since the time of Descartes, humans have insisted that
language exists only in the form of human speech. But even as scientists began discovering the cleverness of dolphin signature whistles and killer whale dialects, and considering the possibility that
these marine mammals might be communicating with unique languages of their own, the emphasis has remained on sound and vocal
learning.

The idea that language should necessarily have to take the form
of human speech and involve vocal learning reflects a "glottocentric" bias, according to Sherman Wilcox, of the Department of Linguistics at the University of New Mexico at Albuquerque. Wilcox
stresses that many members of modern human society communicate with American Sign Language with no need for making or
hearing speech sounds. Wilcox is among the few linguists who have

challenged the basic assumptions about what constitutes a language. The bias was reflected by Giulio Tarra, one of the leading educators of the deaf in the nineteenth century, who believed that deaf people should be taught to make sounds rather than signs. Tarra wrote:

> Gesture is not the true language of man which suits the dignity of his nature. Gesture, instead of addressing the mind, addresses the imagination and the senses. Moreover, it is not and never will be the language of society. . . . Thus, for us it is an absolute necessity to prohibit that language and to replace it with living speech, the only instrument of human thought.

Wilcox's ideas, and those of a few other linguists who have been bold enough to challenge the long-held notion that only human speech is language, have tremendous implications for the perceptions we hold about the conversations taking place in the ocean and in the trees outside our windows. According to Wilcox, "language does not depend on its modality. Language may be expressed across different modalities. Signed languages are not simply gestures. They are richly complex in their internal structure and capable of being combined into fully formed sentences. They also are not signed surrogates of spoken words. Signed languages are natural human languages with distinct grammars unrelated to spoken languages. They are languages."

The pioneer of these new ideas about human language was William C. Stokoe, Jr., a professor emeritus at Gallaudet University at the time of his death in 2000. Stokoe spent much of his long career persuading linguists and society that American Sign Language is a true language, and ensuring that ASL was taught in schools and eventually accepted as a substitute for foreign language requirements in universities. Ironically, he then stirred further controversy by suggesting that human language has its roots in the gestures of primates:

> Communication by a system of gestures is not an exclusively human activity, so that in a broad sense of the term, sign language is as old as the race itself, and its earliest history is equally obscure . . . To take a hypothetical example, a shoulder shrug, which for most speakers accompanied a vocal utterance, might be a movement so slight as to be outside the awareness of most speakers; but to the deaf person, the shrug is unaccompanied by anything perceptible except a predictable set of circumstances and responses; in short, it has definite meaning.

Noam Chomsky was perhaps the most influential linguist of the twentieth century, and his ideas about the origin of human language have had a significant effect on the thinking of the public and animal communication scientists. Chomsky's main theory about the evolution of human language entails the occurrence of a major mutation in humans that led to the development of a universal, innate grammar and syntax. This is what Chomsky had to say in 1972 about the difference between humans and animals:

> When we ask what human language is, we find no striking similarity to animal communication systems. There is nothing useful to be said about behavior or thought at the level of abstraction at which animal and human communication fall together. The examples of animal communication that have been examined to date do share many of the properties of human gestural systems, and it might be reasonable to explore the possibility of direct connection in this case. But human language, it appears, is based on entirely different principles.

Wilcox says he finds Chomsky's words somewhat fanciful. Chomsky seems to be saying that human language and animal communication have no evolutionary bridge, but Chomsky would have no problem agreeing that the natural language of animals that we have discussed throughout this book and the nonverbal communication of humans are connected or even one and the same. For Wilcox, this implies that human gestural communication and language can have no evolutionary link, either.

Says Wilcox: "For Chomsky there are thus two unbridgeable gaps: between animal communication and human language (which for Chomsky is entirely defined by syntax), and between human gesture and human language. Little wonder then that those who follow in Chomsky's tracks must rely on sudden genetic mutations to account for *the whole of linguistic evolution.* Indeed, others believed that syntax must have emerged in one piece, at one time, because of some kind of mutation that reorganized the brain. This does not explain the tight integration between human gesture and human language?"

Stokoe was truly a pioneering thinker: as early as the 1960s, he was arguing for an evolutionary continuity between animal communication and human language, and between human gesture and human language. As Wilcox says, "We don't need a mutational miracle to bridge the gaps; nature, in the form of incremental Darwinian selection, is enough. Stokoe's insight is that in order to see how a continuity view of language evolution could work, we have to correctly identify the raw material on which such selection acts. If that raw material were only acoustic signals . . . we would never have crossed the animal communication to human language gap, and we would not see the intimate link between human gesture and human language."

Vocal learning provides incontrovertible evidence for the origins of human speech, just as it does for birdsong, whale song, signature whistles, and orca dialects. But it does not necessarily lead us to the origin of language. Language may be something other than speech, and it may not be unique to the human species, but to find its roots, we need look no further than the minds of our fellow cognitive creatures.

In 1980 a bonobo named Kanzi was born at the Yerkes Field Station in Georgia. Within 30 minutes of his delivery, the dominant female in his troop, Matata, took Kanzi from his mother and adopted him as her own. When Kanzi was six months old, he and Matata were transferred to the Georgia State University Language Research Center in Atlanta. Sue Savage-Rumbaugh, a primatologist at Georgia State, began working with Matata in an attempt to teach her an artificial language consisting of lexigrams, which are printed symbols, placed on a cardboard keyboard. Savage-Rumbaugh and a team of researchers worked with Matata for nearly two years to teach her the artificial language, in the hope that Matata would learn the symbols and be able to construct simple sentences. Matata seemed incapable of learning, however, and young Kanzi was a constant interruption to the lessons, in which he displayed no interest whatsoever. By the time Kanzi was two and half years old, he was weaned from Matata so that she could be taken back to the field station to be mated. Kanzi searched for her for three days but then gave up. Soon after, he surprised the research team by demonstrating that he knew perfectly how to employ the lexigrams. Even though he had not appeared to pay attention, he had learned everything that the researchers had tried unsuccessfully to teach Matata.

Kanzi's example has important implications for understanding how language is learned. First, the saying You can't teach an old dog new tricks has as much relevance to language as it does to birdsong. This is not to say that Matata could not learn new bonobo vocalizations or behaviors that come naturally to her species, but she had passed a stage at which she might have been socialized to learn something as complex as human language.

A human child reared in isolation, or in the wild without human contact, will not learn how to speak or develop human language, either. The raw materials are there, just as they are in Kanzi, but human language must be taught. On January 8, 1800, a young boy was discovered digging potatoes from a garden in the village of Saint-Sernin in southern France. He was wild in the truest sense, having apparently lived much of his life in the forests like an animal. He did not speak, and he seemed unaware that he was human. The following is a description of him by his caretaker:

> When he is sitting down and even when he is eating, he makes a guttural sound, a low murmur; and he rocks his body from right to left or backwards and forwards, with his head and chin up, his mouth closed and his eyes staring at nothing. In this position he sometimes has spasms, convulsing movements that may indicate that his nervous system has been affected. There is nothing wrong with the boy's five senses, but their order of importance seems to be modified. He relies first on smell, then on taste; his sense of touch comes last. His sight is sharp; his hearing seems to shut out many sounds people around him pay close attention to. Nothing interests him but food and sleep . . . This child is not totally without intelligence, reflection, and reasoning power. However . . . one can only perceive in him animal behaviour.

There have been numerous reports of feral children, but the Wild Boy of Aveyron, as he became known, is one of the best documented. A medical expert today might make a diagnosis of autism, but in this case no one knew how long the boy had been living on his own or how he ended up alone in the forest. Without socialization and exposure to language, the child had no recognizable ability to communicate. Kanzi, on the other hand, is an animal who with exposure to an artificial human language developed quite human qualities, including language.

Kanzi, the bonobo, can construct 650 sentences following proper rules of grammar and syntax. Kanzi learned to use an artifical language by observing lessons given to his mother when he was still an infant.

I pursued the Kanzi story to a linguist at the College of William and Mary in Williamsburg, Virginia. Talbot Taylor, who described himself as a dyed-in-the-wool Cartesianist, at first was entirely skeptical of Kanzi's language abilities. Taylor set out, not unlike Oskar Pfungst, to prove that Kanzi actually had no concept of what he was doing and was most likely reading cues off of his trainers. After spending only a few hours with Kanzi, Taylor was dumbfounded, and he ended up coauthoring *Apes, Language, and the Human Mind* with Savage-Rumbaugh and the philosopher Stuart Shanker. By the time Kanzi was eight years old, he could understand 650 different sentences, among them:

> Tickle Rose with the sparklers.
> Give the doggie [a toy dog] some yogurt.
> Put the toothbrush in the lemonade.
> Take the snake [a plastic snake] outdoors.
> Go scare Matata with the snake.
> Can you pour the ice water in the potty?

Can you take the gorilla [a toy] to the bedroom?
Go get the balloon that's in the microwave.
Can you put the blanket on the doggie?
Can you put the bunny on your hand?
>Make the snake bite Linda.
>Drink the coffee that's hot.
>Drink the iced coffee.
>I want Kanzi to grab Rose.
>Pour the Coke in the lemonade.
>Pour the lemonade in the Coke.

According to Taylor:

The linguist tells us that to understand sentences such as these a hearer must do more than simply know the meanings of the words used. The hearer must also understand various types of grammatical and conceptual relations, comprehend word order patterns, and have some grasp of the creative possibilities of English sentence structure. Yet these requirements are said by the "language gap" theorists only to be achievable by a creature who possesses an autonomous language faculty. However, again, there seems no good reason to attribute possession of such a faculty to Kanzi. What this example reveals is the irrelevance—to the language-origins question—of arguments about whether various animals in the wild or some laboratory-reared chimpanzees do or do not "have language." For, regardless of how one votes on that (fundamentally misleading) question, it is hard to dispute the claim that current research has established that various nonhuman animals manifest at least some cognitive and communicational abilities that the dogma of the "language gap" has long insisted could *only* be the properties of a creature that possesses language. This conclusion should therefore make us question the existence of such a language gap. And if that gap is itself only a conceptual illusion, then the question of the origin of language loses the meaning it has long been taken to have as the explanation of how that crucial gap was crossed.

Taylor proposes that language abilities exist in many species. How a species happens to make use of those abilities accounts for the differences that we consider so important between humans and animals. From a dolphin's perspective, we may seem terribly inept at understanding its perfectly sensible dialogue of whistles, burst-pulses, and click trains. Looking for the roots of language in biological codes in genes may be a fallacy, because purely environmental circumstances and different ecological pressures may give rise to shifting fortunes in the use of language abilities throughout the animal kingdom. For the development of language skills, nurture may be the most important component. Taylor suggests that language skills may come and go many times in a single species.

If a catastrophic event were to wipe out all but the youngest children on earth before they could be taught language by their human parents, they would be no more skilled in human language than the Wild Boy of Aveyron. Their language abilities would still be intact, but the skills would be lost. It is possible, according to Taylor, that bonobos or chimpanzees might have had more advanced language skills at another time in history and lost them in a catastrophic event. Language skills are acquired through the development of culture. Human skills have advanced because our species found a way of holding on—so far, anyway—to our language tradition. Perhaps the origin of human language has seemed such a mystery and required such fantastic explanations because we have ignored its roots in other natural communication systems.

In the beginning of this journey through animal talk, we looked at the biblical phrase "Now the whole earth had one language and few words." The sinews of this common language extend into every creature in the animal kingdom. Some animals more than others have evolved layers of complexity on top of this one language, but there has never been a gap between humans and animals. Our bodies come from their bodies, and our human language stems from their communications systems. Charles Snowdon at the University of Wisconsin takes a more conservative view, but not by much: "The finding of some rudiments of language-like phenomenon in natural communication, such as simple grammars, rudimentary symbolization, similar perceptual systems, along with the potential exhibited by some of the great apes, indicates that many of the components of human linguistic ability have appeared at different times in evolution; but it is only with human beings that the components that define our linguistic abilities have come together in one species."

There is no doubt that human language is a remarkable achievement that has developed and matured in our culture for at least tens of thousands of years. But does language necessarily have to be a human invention? Many species have been shown to possess a basic language capability through the attempts to teach artificial languages to species such as dolphins, parrots, and nonhuman primates like Kanzi. Why would these animals possess a fundamental capability for grammar and syntax if they did not make use of them somehow in their own species-specific ways to communicate? As more scientists overcome the biases that have influenced the designs of animal communication studies for the past century and move away from the longer-held notion that if an animal can't use language like a human it must not have language, perhaps then we may begin to appreciate the possibility that more than one way—our way—exists to conduct a conversation. The minds of animals and their lives may be simpler than ours, but that does not mean they are not thinking, feeling, making decisions, and communicating in ways that are more sophisticated than we have imagined.

As I said before, all creatures spend much of their time talking about sex, real estate, who's boss, and what's for dinner. And we all tend to express our aggression, surprise, affection, and lust in quite similar ways. Sure, some of us have amazing powers of speech, while others use clicks and sonar, or wave their hands, or flash brilliant feathers, or bark, or chirp, or whinny. But maybe Darwin was right more than 150 years ago when he described our differences as ones of degree and not of kind. The different species on this planet may sound different and look different, but all are members of the same extended family, living here for the same essential reasons. And when we open our mouths, it's all animal talk.

Notes

The number preceding note is the page number to which it refers.

One: A Walk in the Park: Toward a Universal Language

5–6. Definitions of communication. Jack Bradbury and Sandra Vehrencamp. *Principles of Animal Communication.* Sunderland, MA: Sinauer Associates, 1998. Edward O. Wilson. *Sociobiology, The New Synthesis,* 25th anniversary edition, Cambridge, MA: The Belknap Press of Harvard University Press, 2000. Lesley Rogers and Gisela Kaplan. *Songs, Roars and Rituals: Communication in Birds, Mammals and other Animals.* Cambridge, MA: Harvard University Press, 2000.

7. Bacterial communication and quorum sensing. Steven Schultz. *Princeton Weekly Bulletin.* March 29, 1999. Yi-Hu Dong, Lian-Hui Wang, Jin-Ling Xu, Hai-Bao Zhang, Xi-Fen Zhang, Lian-Hui Zhang. "Quenching Quorum-Sensing-Dependent Bacterial Infection by an *N*-acyl Homoserine Lactonase." *Nature* 411 (June 14, 2001):813–817. Paul Watnick and Roberto Kolter. "Biofilm, City of Microbes." *Journal of Bacteriology* 182 (May 2000): 2675–2679.

9. The information model versus behavioral manipulation as the underlying purpose of communication. Richard Dawkins and John R. Krebs. "Animal Signals: Information or Manipulation?" in *Behavioral Ecology.* Oxford: Blackwell, 1978. Richard Dawkins. *The Selfish Gene.* New York: Oxford University Press, 1989. Eugene Morton and Donald Owings. *Animal Vocal Communication, A New Approach.* Cambridge: Cambridge University Press, 1998.

11. Not responding to a signal is still communication. Harold Gouzoules, personal communication.

13. Electrical fields as signals. Philip K. Stoddard. "Predation Enhances Complexity in the Evolution of Electric Fish Signals." *Nature* 400 (1999): 254–256.

15. Elephant Viagra. L. E. L. Rasmussen, et al. "Insect Pheromone in Elephants." *Nature* 379 (1996):684. Josef Lazar, David R. Greenwood, L. E. L. Rasmussen, and Glenn Prestwich. "Molecular and Functional Characterization of an Odorant Binding Protein of the Asian Elephant, Elephas Maximus: Implications for the Role of Lipoclains in Mammalian Olfaction." *Biochemistry* 41 (2002):11786–11794.

16. Sound channel used for long distance calls by whales. Sound Fixing and Ranging (SOFAR) Channel description provided by the National Oceanic and Atmospheric Administration.

17. Quote about Ken Norris by Daniel P. Costa. Henry Fountain. "Kenneth Norris, 74. Pioneer in Study of Marine Mammals" (obit). *New York Times* (Sunday, Aug. 23, 1998).

18–19. Senses and their role in communication. Bradbury and Vehrencamp, ibid.

20–21. Low-frequency signals of elephants. Katherine B. Payne, William R. Langbauer Jr., and Elizabeth M. Thomas. "Infrasonic Calls of the Asian Elephant." *Behavioral Ecology and Sociobiology* 18 (1986):297–301. Michael Garstang, et al. "Meteorology and Elephant Infrasound at Etosha National Park, Namibia." *Journal of the Acoustical Society of America* 101, no. 3 (Mar. 1997): 1710–1717.

21. Technology for sensing marine mammal communication. Peter Tyack, personal communication.

22. Borneo tree frog. Bjorn Lardner and M. bin Lakim. "Tree-Hole Frogs Exploit Resonance Effects." *Nature* 420 (Dec. 5, 2002):475.

24. The orchestra in the rainforest. Bernard L. Krause, PhD. "The Niche Hypothesis: A Virtual Symphony of Animal Sounds, the Origins of Musical Expression and the Health of Habitats." *The Soundscape Newsletter,* no. 6 (June 6, 1993):4–6.

Two: Of Ravens and Robots

34–35. Tale of a raven and mountain lion. Bernd Heinrich, personal communication.

37–38. Betty the Crow. Alex A. S. Weir, Jackie Chappell, and Alex Kacelnik. "Shaping of Hooks in New Caledonian Crows." *Science* 297 (Aug. 9, 2002):981. Personal reporting.

39–40. Mystery of the blue tits and milk bottle caps. James Gould and Carol Grant. *The Animal Mind.* New York: Scientific American Library, 1999. This book provided a rich source of information on animal cognition, mate selection, and anecdotes. I conducted numerous interviews with James Gould during my research for this book.

41–42. The trouble with Descartes. The discussion of Descartes' *Discours de la Methode* and the creation of the divide between humans and animals are based on writings by philosopher Stuart G. Shanker of York University in Toronto, coauthor of *Apes, Language, and the Human Mind*, with Sue Savage-Rumbaugh (Georgia State University) and linguist Talbot Taylor (College of William and Mary). In Descartes' words, "By these two methods we may also recognize the difference that exists between men and brutes. For it is a very remarkable fact that there are none so depraved and stupid, without even excepting idiots, that they cannot arrange different words together, forming of them a statement by which they make known their thoughts; while, on the other hand, there is no other animal, however perfect and fortunately circumstanced it may be, which can do the same. It is not the want of organs that brings this to pass, for it is evident that magpies and parrots are able to utter words just like ourselves, and yet they cannot speak as we do, that is, so as to give evidence that they think of what they say. On the other hand, men who, being born deaf and dumb, are in the same degree, or even more than the brutes, destitute of the organs which serve the others for talking, are in the habit of themselves inventing certain signs by which they make themselves understood by those who, being usually in their company, have leisure to learn their language. And this does not merely show that the brutes have less reason than men, but that they have none at all, since it is clear that very little is required in order to be able to talk. And when we notice the inequality that exists between animals of the same species, as well as between men, and observe that some are more capable of receiving instruction than others, it is not credible that a monkey or a parrot, selected as the most perfect of its species, should not in these matters equal the stupidest child to be found, or at least a child whose mind is clouded, unless in the case of the brute the soul were of an entirely different nature from ours. And we ought not to confound speech with natural movements which betray passions and may be imitated by machines as well as be manifested by animals; nor must we think, as did some of the ancients, that brutes talk, although we do not understand their language. For if this were true, since they have many organs which are allied to our own, they could communicate their thoughts to us just as easily as to those of their own race. It is also a very remarkable fact that although there are many animals which exhibit more dexterity than we do in some of their actions, we at the same time observe that they do not manifest any dexterity at all in many others. Hence the fact that they do better than we do, does not prove that they are endowed with mind, for in this case they would have more reason than any of us, and would surpass us in all other things. It rather shows that they have no reason at all, and that it is nature which acts in them according to the disposition of their organs, just as a clock, which is only composed of wheels and weights is able to tell the hours and measure the time more correctly than we can do with all our wisdom."

47. Reductio ad absurdum. E. O. Wilson, on interpretations of Morgan's Canon. *Sociobiology: The New Synthesis, 25th Anniversary Edition.* Cambridge: The Belknap Press of Harvard University Press, 2000: 176–200.

47–50. Morgan is misinterpreted. Roger K. Thomas. "Lloyd Morgan's Canon: A History of Misrepresentation." *History & Theory of Psychology Eprint Archive* (2001). Lloyd C. Morgan. *Animal Mind.* New York: Longmans, Green & Co., 1930.

50–53. Clever Hans. Oskar Pfungst. *Clever Hans (The Horse of Mr. von Osten)* Edited by Robert H. Wozniak. Reprint, Bristol, UK: 1907; English 1911. Thoemmes Press, 1998.

54. Window into the animal mind. Donald R. Griffin, *Animal Minds.* Chicago: University of Chicago Press, 1992.

55–57. Alarm calls. Griffin, ibid. Robert M. Seyfarth and Dorothy L. Cheney, "The Ontogeny of Vervet Monkey Alarm Calling Behavior: A Preliminary Report." *Zeitschrift für Tierpsychologie* 54 (1980):37–56.

60. Deceptive alarm call in macaques. Harold Gouzoules, personal communication.

61. The mental tool kit. Marc D. Hauser. *Wild Minds, What Animals Really Think.* New York: Henry Holt and Company, 2000.

Three: The Show Must Go On

64–74. Midway Island and albatross behavior. Information regarding courtship among juvenile albatrosses is based on personal research observations and interviews with scientists at Midway who were working with the U.S. Fish and Wildlife Service and the Oceanic Society, San Francisco. The Oceanic Society is a non-profit group that conducts field research with scientists and non-scientist volunteers at Midway and areas in the Caribbean and Amazon. The group was extremely helpful in providing me with accommodations and transportation to the remote island. One of the more important activities of the society at Midway is an attempt to protect the endangered monk seal, which raises its young there in shallow coves.

70–74. Learning and cultural transmission of behavior. Lee Dugatkin. *The Imitation Factor, Evolution Beyond the Gene.* New York: Free Press, 2000, and personal communication. Frans de Waal. *The Ape and the Sushi Master.* New York: Basic Books, 2001, and personal communication. Both Dugatkin and de Waal were extremely helpful in explaining cultural transmission of behavior in animals. I did not devote many pages to animal culture, but it is another important area of research that is bridging the gap between humans and animals. Few scientists today doubt that animals learn new behaviors in their lifetimes and pass those behaviors along to their offspring. Bobbi S. Low. *Why Sex Matters: A Darwinian Look at Human Behavior.* Princeton, NJ: Princeton University Press, 2000. Foraging behavior in ground squirrels is considered to be 60 percent inherited and 40 percent learned by offspring from their mothers.

75–78. Theory of evolution. Charles Darwin. *Origin of Species.* 1859.

74. Ambient sound of rain forest. Bernard Krause, ibid.

78–81. Genes and evolution. Personal reporting on the Human Genome Project. I have covered the Human Genome Project for *USA Today* since the project's inception in the late 1980s.

81–83. Cuddle chemicals. Tom Insel, Emory University, personal communication. I have written several articles for the newspaper on studies Insel conducted throughout the 1990s. Monogamy and social bonding in humans and animals are dependent on hormones such as oxytocin and vasopressin. More recent research on the physiology of behavior is led by Charles T. Snowdon of the University of Wisconsin.

Four: One Language and Few Words

85–94. The motivation structural rule. Eugene S. Morton. "On the Occurrence and Significance of Motivation Structural Rules in Some Bird and Mammal Sounds." *American Naturalist* 111 (1977):855–869. Donald H. Owings and Eugene S. Morton, *Animal Vocal Communication: A New Approach.* New York: Cambridge University Press, 1998. I spent many hours discussing the MS rule with Morton, who taught me much about the basics of animal communication. Some of the scientists I spoke with about the motivation structural rule scoffed and downplayed its significance. One called it "old ethology" and no longer of any interest. The subject receives only one paragraph in *Principles of Animal Communication.* It may not be important to scientists who are studying specific species, but it struck me during my research for this book that it was vital to explaining how different species understand each other. I was encouraged by Peter Tyack in early discussion to pursue this line of thinking. I discovered Morton's paper buried in the treasure of files at Tyack's lab. Morton subsequently spent many hours helping me understand the basics of communication and how the rule fit into the grander picture.

93. Lupey, the wolf. John Fentress discussed wolf behavior and communication at a Smithsonian Associates lecture held in Washington, D.C., 2000.

98. Piloerection in a frightened chimpanzee. Jane Goodall, personal communication.

99. Principle of Antithesis. Charles Darwin. *Expression of the Emotions in Man and Animals.* 1872.

99–104. Pitch and dominance effects of voices. Stanford W. Gregory, Jr. and Timothy J. Gallagher. "Spectral Analysis of Candidates' Nonverbal Vocal Communication: Predicting U.S. Presidential Election Outcomes." *Social Psychology Quarterly* 65, no. 3 (2002):298–308. Ray L. Birdwhistle. "The Language of the Body: The Natural Environment of Words." In *Human Communication: Theoretical Explorations.* Edited by Albert Silverstein. New York: Wiley, 1974.

102. Right hemisphere of brain and nonverbal communication. Julian Keenan, personal communication, 2003.

104–5. Communication between mothers and infants. Anne Fernald. "Human Maternal Vocalizations to Infants as Biologically Relevant Signals: An Evolutionary Perspective." In *The Adapted Mind, Evolutionary Psychology and the Generation of Culture.* Edited by Jerome H. Barkow, Leda Cosmides, and John Tooby. New York: Oxford University Press, 1992.

107–8. Universal voice qualities. David Givens, Center for Nonverbal Studies, Spokane, Washington, Personal communication.

108. Differences in the sexes' expression and perception of nonverbal signals. M. A. Griffin, D. McGahee, and J. Slate. "Gender Differences in Nonverbal Communication." Valdosta State University, 1999: www.bvte.ecu.edu/ ACBMEC/p1999/Griffin.htm. J. K. Burgoon, D. B. Buller, and W. G. Woodall. *Nonverbal Communication: The Unspoken Dialogue,* 2nd ed. New York: McGraw-Hill, 1996. M. S. Hanna, and G. L. Wilson. *Communicating in Business and Professional Settings,* 4th ed. New York: McGraw-Hill, 1998. D. K. Ivy, and P. Backlund. *Exploring Genderspeak.* New York: McGraw-Hill, 1994.

Five: The Chemistry of Love

112–15. Bacterial signaling. In addition to the references listed in chapter one: Bonnie Bassler. "Tiny Conspiracies, Cell-to-cell Communication Allows Bacteria to Coordinate Their Activity." *Natural History* (May 2001). J. W. Costerton, P. S. Stewart, and E. P. Greenberg. "Bacterial Biofilms: A Common Cause of Persistent Infections." *Science* 284, no. 5418 (May 1999):1318–1322. W. J. Gould. *Wonderful Life: The Burgess Shale and the Nature of History.* New York: W. W. Norton, 1989. P. K. Singh, et al. "Quorum-Sensing Signals Indicate That Cystic Fibrosis Lungs Are Infected with Bacterial Biofilms." *Nature* 407 (2000):762–764.

117–18. Fireflies' flash patterns. M. A. Branham, and M. D. Greenfield. "Flashing Males Win Mate Success." *Nature* 318 (1996):745–746.

118–20. Femme fatales. Thomas Eisner, Cornell University, personal communication. *Proceedings of the National Academy of Sciences* 94 (Sept. 2, 1997): 9723–9728.

122–23. Alpha female monopolizes reproduction. T. Mannin, et al. "Pretender Punishment Induced by Chemical Signalling in a Queenless Ant." *Nature* 419 (Sept. 2002):61–65.

127. Dormitory effect. M. K. McClintock. "Menstrual Synchrony and Suppression." *Nature* 291 (1971):244–245. M. K. McClintock. "Pheromonal Regulation of the Ovarian Cycle: Enhancement, Suppression and Synchrony." In J. G. Vandenbergh (ed.), *Pheromones and Reproduction in Mammals.* New York: Academic Press, 1983.

128. Men like the way women smell during peak ovulation. *BBC News,* Tuesday, April 3, 2001.

129. Women sensitive to musky smells of men. J. J. Cowley, and B. W. L. Brooksbank, "Human Exposure to Putative Pheromones and Changes in Aspects of Social Behavior." *Journal of Steroid Biochemistry and Molecular Biology* 39 (1991):647–659.

130–32. MHC genes. K. Yamazaki, et al. "Parent-Progeny Recognition as a Function of MHC Odortype Identity." *Proceedings of the National Academy of Sciences* 19 (Sept. 12, 2000):10500–2. R. M. Younger, et al. "Characterization of Clustered MHC-Linked Olfactory Receptor Genes in Human and Mouse." *Genome Research* 11, no. 4 (April 2001):519–530. Y. Kobayashi and Larry R.

Fisher, *Pheromones: The Smell of Beauty,* International English Center, University of Colorado (Published Thesis, 1997).

129–30. VNO. *Seeing, Hearing and Smelling the World.* Report from the Howard Hughes Medical Institute: www.hhmi.org/senses.

131. Female stickleback fish prefer the scent of males associated with MHC genes most different from their own. Thorsten Reusch."Female Stickleback Count Alleles: A New Strategy of Sexual Selection Explaining MHC Polymorphism." *Nature* 414 (2001):300–302. Erica Klarreich. "My, You Smell Diverse." *Nature Science Update* (Nov. 15, 2001).

131–32. MHC typing and perfume. "The Sweet Smell of the Immune System." John Whitfield, *Nature Science Update* (March 7, 2001).

135. Handicap Principle. Amot Zahavi. "Mate Selection: A Selection for a Handicap." *Journal of Theoretical Biology* 53 (1975):205–214.

136–37. General rules. Bradbury and Vehrencamp, ibid.

Six: Songs and Shouts

138–41. To a female midshipman's ears, the loud low hum of the love song must sound like sweet music. M. S. Weeg, R. R. Fay, and A. H. Bass. "Directionality and Frequency Tuning of Primary Saccular Afferents of the Sonic Fish Porichthys Notatus (Midshipman)." *Journal of Comparative Physiology* A 188 (2002):631–641. J. L. Goodson and A. H. Bass. "Forebrain and Midbrain Vocal-Acoustic Complexes: Intraconnectivity and Descending Vocal Motor Pathways." *The Journal of Comparative Neurology* 448 (2002):298–321. J. R. McKibben and A. H. Bass. "Peripheral Encoding of Behaviorally Relevant Acoustic Signals in a Vocal Fish: Harmonic and Beat Stimuli." *Journal of Comparative Physiology* A 187 (2001):271–285.

141–45. Songs of the humpback whale. Roger Payne, personal communication. Peter Tyack, personal communication. Janet Mann, et al. *Cetacean Societies: Field Studies of Dolphins and Whales.* Chicago: University of Chicago Press, 2000. I could list dozens of other references for studies of humpback whale songs and behavior, but for one-stop shopping, most of these can be found at the back of *Cetacean Societies.*

149–54. Information on song learning in birds is based primarily on extensive interviews with Eugene Morton and Jack Bradbury, and much late-night reading of their textbooks *Animal Vocal Communication: A New Approach* and *Principles of Animal Communication.* Peter Tyack supplied me with dozens of scientific papers on the same topic. Among those cited: Peter Marler. "Song Learning: The Interface between Behaviour and Neuroethology." *Philosophical Transactions of The Royal Society of London: Biological Sciences* 329 (1990):109–114. Mary Sue Waser and Peter Marler. "Song Learning in Canaries." *Journal of Comparative and Physiological Psychology* 91 (1977). F. Nottebohm, T. Stokes, and C. M. Leonard. "Central Control of Song in the Canary." *Journal of Comparative Neurology* 165 (1966):457–486. S. W. Bottjer, E. A. Miesner, and A. P. Arnold.

"Forebrain Lesions Disrupt Development but not Maintenance of Song in Passerine Birds." *Science* 224 (1984):901–903. D. E. Kroodsma and B. E. Byers. "The Function(s) of Bird Song." *American Zoology* 31 (1991):318–328. P. J. B. Slater, L. A. Eales, and N. S. Clayton. "Song Learning in Zebra Finches (*Taeniopygia guttata*):Progress and Prospects." *Advances in the Study of Behavior,* vol. 18. Edited by J. S. Rosenblatt, C. Beer, M. C. Busnel, and P. J. B., Slater. New York: Harcourt, 1998. P. Slater. "Bird Song Learning: Causes and Consequences." *Ethology, Ecology, and Evolution* 1 (1989):19–46.

150. Young zebra finches dream about the songs they have heard during the day. D. Margoliash. "Do Sleeping Birds Sing? Population Coding and Learning in the Bird Song System?" *Progress in Brain Research,* vol. 130 (Jan. 1, 2001):319–31. A. S. Dave, and D. Margoliash. "Song Replay During Sleep and Computational Rules for Sensorimotor Vocal Learning." *Science* 290 (5492) (Oct. 27, 2000):812–6.

150. Female songbird's discerning nature. D. J. Mennill, L. M. Ratcliffe, and P. T. Boag. "Female Eavesdropping on Male Song Contests in Songbirds." *Science* 296 (2002):873.

152–53. Male white-crowned sparrow. P. Marler. "An Ethological Theory of the Origin of Vocal Learning." *Annals of the New York Academy of Sciences* 280 (1976):386–395.

154. Decline of songbirds. Su Engstrand. "Effect of Noise on Blackbird Song." Bird and Mammal Sound Communication Group, University of St. Andrews, abstract: www.st-andrews.ac.uk/~bmscg/turdus.htm.

155–57. Duetting in gibbons. T. Geissmann. *Evolution of Communication in Gibbons (Hylobatidae).* Ph.D. thesis, Anthropological Institute, Philosoph. Faculty II, Zürich University, 1993. 374 pp.

158. The matriarch (elephant), like our grandmothers, possesses valuable knowledge that she can communicate. K. McComb, C. Moss, et al. "Matriarchs as Repositories of Social Knowledge in African Elephants." *Science* 292 (April 20, 2001):491.

159–64. Elephant communication. Joyce H. Poole, et al. "The Social Contexts of Some Very Low Frequency Calls of African Elephants." *Journal of Ecology and Sociobiology* 22 (1988):385–392. Katherine Payne, et al. "Infrasonic Calls of the Asian Elephant." *Journal of Behavioral Ecology and Sociobiology.* 18 (1986):297–301. I also learned much about elephant communication and behavior from *Silent Thunder, In the Presence of Elephants* by Katy Payne (Simon & Schuster, 1998), and *Elephant Memories, Thirteen Years in the Life of an Elephant Family,* by Cynthia Moss (Fawcett Columbine, Ballantine Books, 1988). Both Payne and Moss spent time during interviews for this book to explain their work in more detail.

163–64. Feet as a sensory organ. Caitlin O'Connell-Rodwell. "Seismic Properties of Asian Elephants (Elephas maximus) Vocalizations and Locomotion." *Journal of the American Acoustical Society* 108 (6)(Dec. 2000):3066–72.

O'Connell-Rodwell began her research in the area of stridulations of insects, another important area of animal communication that is absent from this book. If I'd had more time, I would have included another chapter on vibrations as signals and the use of the tactile sense to detect them. O'Connell-Rodwell continues her research with elephants and other large mammals, as well as with insects. I had an opportunity to interview her for this book and learn about an entire field of research that I'd completely overlooked.

Seven: Flash and Dance

166–70. General information on visual signals is obtained from *Principles of Animal Communication* by Bradbury and Vehrencamp, and from personal communications with Bradbury.

170–71. Diets provide carotinoids for colorful feathers. K. J. Navara and G. E. Hill, "Dietary Carotinoids Pigments and Immune Function in a Songbird with Extensive Carotinoids-Based Plumage Coloration." *Behavioral Ecology,* in press.

171–72. Disruptive selection in male lazuli buntings. E. Greene, B. E. Lyon, et al. "Disruptive Sexual Selection for Plumage Coloration in a Passerine Bird." *Nature* 407 (Oct. 26, 2000):1000. T. Slagsvold. "Why Are Some Males Dull?" *Nature* 407 (Oct. 26, 2000):955.

173–74. Intention movements. Bradbury and Vehrencamp, ibid.

175–76. Common steps. James W. Davis and Whitman A. Richards, "Relating Categories of Intentional Animal Motions." Ohio State University, Dept. of Computer and Information Science Technical Report, OSU-CISRC-11/00-TR25 (Nov. 2000).

179–80. Lekking and sage grouse. Bradbury and Vehrencamp, ibid.

181–86. Bowerbirds. Paul A. Johnsgard. *Arena Birds: Sexual Selection and Behavior.* Washington, D.C.: Smithsonian Institution Press, 1994. Donald R. Griffin, *Animal Minds.* Chicago: University of Chicago Press, 1992. J. Albert C. Uy, Gail L. Patricelli, and Gerald Borgia. "Dynamic Mate-Searching Tactic Allows Female Satin Bowerbirds (Ptilonorhynchus violaceus) to Reduce Searching." Proceedings of the Royal Society of London: Biological Sciences 267 (2000):251–256. J. Albert C. Uy, Gail L. Patricelli, and Gerald Borgia. "Loss of Preferred Mates Forces Female Satin Bowerbirds (Ptilonorhynchus violaceus) to Increase Mate Searching." Proceedings of the Royal Society of London 268 (2001):633–638. "Bachelor Pads for the Birds." James Cook University media release, Oct. 10, 2000. Gail L. Patricelli, J. Albert C. Uy, Gregory Walsh, and Gerald Borgia. "Sexual Selection: Male Displays Adjusted to Female's Response." *Nature* 415 (2002):279–280.

185–89. Visual signals for attracting mates in primate species. Barbara Smuts, et al. *Primate Societies.* Chicago: University of Chicago Press, 1987.

189–90. Bobbi S. Low, *Why Sex Matters: A Darwinian Look at Human*

Behavior. Princeton: Princeton University Press, 2000. Low was helpful in pointing out areas of her book that are related to mate attraction in primates and other species. Her book is incredibly insightful and worth reading.

189-91. Visual mating signals in humans. David Givens, Center for Nonverbal Studies, personal communication. Givens has a remarkable website devoted to nonverbal communication. Men and women who study the pages on visual mate attraction signals will have an advantage over their potential sexual partners. People are usually not aware of the signals they display when communicating in public because we have been trained to place so much emphasis on speech. Human nonverbal communication is a small but extremely active field of research that has been overlooked by many of the people studying animal communication. It is fair to say the social scientists and psychologists who study nonverbal communication in humans are more informed about animal communication than animal communication scientists are about human nonverbal communication.

Eight: Our House

I owe almost every source of information in this chapter to Jack Bradbury and Sandra Vehrencamp. By the time I reached this stage of reporting for this book, the backing had become unglued on my copy of *Principles of Communication.* The book is especially helpful for understanding territorial behavior. One disappointment is that I was unable to convey the importance of the use of game theory to explain territorial behaviors. I did such a poor job of explaining it in an early draft that Bradbury asked me to rewrite it and my editor asked me finally to just remove it. For the general public it is perhaps not so important. But for anyone interested in studying animal communication it is a critical component of research. Game theory allows one to set up various conditions and then play them out until both sides reach a stalemate or balance known as the *evolutionarily stable strategy.* Then a scientist can compare the game and its outcome with field observations. These games remind me of the types of war games played by generals. Fully a third of *Principles of Communication* is devoted to game theory. I have used the conclusions arrived at from various games and left out how they were reached.

196–98. Defending sunspots; the speckled wood butterfly. N. B. Davies. "Territorial Defense in the Speckled Wood Butterfly (Parage aegeria): The Resident Always Wins." *Animal Behavior* 26 (1978):138-147. Of all the anecdotes I have come across during the research for the book, this one is my favorite. Who would ever guess that butterflies would be jousting in the forest over a temporary splash of sunlight?

198–99. Songbird territoriality. Eugene Morton, personal communication. Jack Bradbury, personal communication.

199–200. Rules of territorial behavior. Bradbury and Vehrencamp, ibid.

204–5. Wild horses and territorial behavior. Ginger Katherines, personal communication. Katherines is a documentary filmmaker who has spent years studying wild horses in the western United States. One of her passions is saving horses from annual roundups and slaughters conducted by the federal government.

206. Chimpanzee patrols. Jane Goodall, personal communication.

206–8. Magpies and Zoe. Eleanor Brown. "What Birds with Complex Social Relationships Can Tell Us About Vocal Learning: Vocal Sharing in Avian Groups. *Social Influences on Vocal Development.* Edited by Charles T. Snowdon and Martine Hausberger. New York: Cambridge University Press, 1997.

Nine: All in the Family

215. The story of marine mammal research opens in 1953. W. N. Kellogg, Robert Kohler, and H. N. Morris. "Porpoise Sounds as Sonar Signals." *Science* 117 (March 6, 1953):239–243.

219. Bubble streams and bubble nets. Janet Mann, et al., *Cetacean Societies.* Peter Tyack. Personal communication.

220–21. Dolphins live in a complex, large society. R. C. Connor. et al. *Cetacean Societies.* R. C. Connor, R. A. Smolker, and A. F. Richards. "Two Levels of Alliance Formation Among Male Bottlenose Dolphins (*Tursiops* sp.)." *Proceedings of the Natural Academy of Sciences* 89 (1992):987–990. S. Leatherwood and R. R. Reeves. *The Bottlenose Dolphin.* San Diego: Academic Press, 1998. J. Mann and B. B. Smuts. "Natal attraction; Allomaternal Care and Mother-Infant Separations in Wild Bottlenose Dolphins." *Animal Behavior* 55 (1998):1097–1113. L. S. Sayigh, P. L. Tyack, R. S. Wells, and M. D. Scott. "Signature Whistles of Free-Ranging Bottlenose Dolphins *Tursiops truncates:* Stability and Mother-Offspring Comparisons." *Behavioral Ecology and Sociobiology* 26 (1990):247–260. S. H. Shane, R. S. Wells, and B. Würsig. "Ecology, Behavior and Social Organization of the Bottlenose Dolphin: A Review." *Marine Mammal Science* 2, no. 1 (1986):34–63.

221–23. Signature whistles in dolphins. Vincent Janik and Peter Tyack, personal communication. R. A. Smolker, J. Mann, and B. B. Smuts. "Use of Signature Whistles During Separations and Reunions by Wild Bottlenose Dolphin Mothers and Infants." *Behavioural Ecology and Sociobiology* 33 (1993):393–402. V. M. Janik, G. Dehnhardt, and D. Todt. "Signature Whistle Variations in a Bottlenose Dolphin, *Tursiops Truncatus.*" *Behavioural Ecology and Sociobiology* 35 (1994):243–248. D. Reiss and B. McCowan. "Spontaneous vocal mimicry and production by bottlenose dolphins (*Tursiops truncatus*): Evidence for Vocal Learning." *Journal of Comparative Psychology* 107 (1993): 301–312. L. S. Sayigh, et al. "Sex Difference in Signature Whistle Production of Free-Ranging Bottlenose Dolphins." *Behavioural Ecology and Sociobiology* 36 (1995):171–177. L. S. Sayigh, et al. "Individual Recognition in Wild Bottlenose Dolphins: A Field Test Using Playback Experiments." *Animal Behaviour* 57 (1998):41–50. P. L. Tyack. "Development and Social Functions of Signature Whistles in Bottlenose Dolphins *Tursiops truncates.*" *Bioacoustics* 8 (1997):21–46.

J. L. Miksis, P. L. Tyack, and J. R. Buck. "Captive Dolphins, *Tursiops truncates*, Develop Signature Whistles That Match Acoustic Features of Human-Made Model Sounds." *Journal of the Acoustical Society of America* 112 (2002):728–739.

222-24. Vocal learning. Vincent Janik. "Origins and Implications of Vocal Learning in Bottlenose Dolphins." In *Mammalian Social Learning: Comparative and Ecological Perspectives.* Edited by H. O. Box and K. R. Gibson. Cambridge: Cambridge University Press, 1999. Vincent Janick and P. J. B. Slater. "Vocal Learning in Mammals." *Advances in the Study of Behavior* 26 (1997):59–99.

226. Low-frequency pop; an aggressive signal to females from males. R. C. Connor and R. A. Smolker. "'Pop' Goes the Dolphin: A Vocalization Male Bottlenose Dolphins Produce During Consortship." *Behaviour* 133 (1996):643–662. R. C. Connor, M. R. Heithaus, and L. M. Barre. "Superalliance of Bottlenose Dolphins." *Nature* 397 (Feb. 1999):571–572. R. C. Connor, A. F. Richards, R. A. Smolker, and J. Mann. "Patterns of Female Attractiveness in Indian Ocean Bottlenose Dolphin." *Behaviour* 133 (1996):37–69.

227–28. Threat postures in bottlenose dolphin. *Cetacean Societies.*

228–34. The killer whale. L. Rendell and H. Whitehead, "Culture in Whales and Dolphins." *Behavioral and Brain Sciences* 24, no. 2 (2001):309–324.

234–35. Learning to recognize the dinner bell. V. B. Deecke, P. J. B. Slater, and J. K. B. Ford. "Selective Habituation Shapes Acoustic Predator Recognition in Harbour Seals." *Nature* 420 (2002):171–173.

235–37. Hoover the talking seal. Personal communication with staff at New England Aquarium and old news releases provided by the media relations department.

238. Vocal learning has evolved to coordinate daily social calendars, maintain affiliations. Eleanor D. Brown and Susan M. Farabaugh. "What Birds with Complex Social Relationships Can Tell Us about Vocal Learning: Vocal Sharing in Avian Groups." In *Social Influences On Vocal Development.* Edited by Charles T. Snowdon and Martine Hausberger. New York: Cambridge University Press, 1997.

239. Evidence for vocal learning in chimpanzees. Andrew J. Marshall, Richard W. Wrangham, and Adam Clark Arcadi. "Does Learning Affect the Structure of Vocalizations in Chimpanzees?" *Animal Behavior* 58 (1999):825–830.

240–41. What constitutes a language? I spent many hours talking about language with Sherman Wilcox and many more poring through his publications. His ideas inspired me to tread into this dangerous territory and attempt to link communication with language. S. Wilcox. "The Ritualization and Invention of Language." In *The Origins of Language: What Nonhuman Primates Can Tell Us.* Edited by B. King. Santa Fe: SAR Press, 1999. S. Wilcox, "Deafness and Sign Language Instruction." In *Concise Encyclopedia of Educational Linguistics.* Edited by Bernard Spolsky. London: Elsivier, 1999. S. Wilcox and P. Wilcox. *Learning to See: American Sign Language as a Second Language,* 2nd edition. Washington, D.C.: Gallaudet University Press, 1996. D. F. Armstrong, W. C. Stokoe, and S. E. Wilcox. *Gesture and the Nature of Language.* New York:

Cambridge University Press, 1995. S. Wilcox. "Structural Generativity, Meaning Generation, and the Origins of Language." Commentary on T. Crow, "The Genetic Origins of Language." *Psychology of Cognition* 17 (1998):1215–1220.

244–48. Kanzi. Talbot Taylor, Stuart G. Shanker, and Sue Savage-Rumbaugh, *Apes, Language and the Human Mind.* New York: Oxford University Press, 1998. I spent numerous hours discussing language with Taylor and first met him in 1998. The ideas of Taylor and Wilcox are, in my opinion, the most logical of any that I have heard regarding language and its origins. The following are citations of material which Taylor either provided or I obtained when he wasn't looking. "Ape Linguistics: Is Kanzi a Cartesian?" (with S. G. Shanker). In *Linguistic Historiography 1996.* Edited by D. Cram *et al.* Amsterdam: J. Benjamins, in press. "Apes with Language." (with S. Savage-Rumbaugh and S. G. Shanker). In *Critical Quarterly* 38, no. 3 (1996):45–57. *Landmarks in Linguistic Thought: The Western Tradition from Socrates to Saussure,* 2nd expanded edition (with R. Harris). London: Routledge, 1997 (first edition, 1989). "The Anthropomorphic and the Sceptical." *Language & Communication,* special issue on primate communication (B. J. King, guest editor), vol. 14, no. 1 (1994):115–127. "Why We Need a Theory of Language." In *Linguistics and Philosophy: The Controversial Interface.* Edited by Rom Harré and Roy Harris. Oxford: Pergamon, 1993.

INDEX

Page numbers in *italics* refer to illustrations.

Abel (crow), 37–38
Aborigines, Australian, 164, *181*
acoustic tag, digital, 21, 146–47, *148*
acquired characteristics, inheritance
    of, 77
adaptation, 72–75, 76, 82
Afghanistan, war in, 94–95
aggressive displays:
    enlargement in, 26, 94–95, 97, 98
    submission in, 98–99
    see also territoriality
agoutis, 123, 168
alarm (distress) calls, of primates, 12,
    28, 29, 35, 55–61, *56*
albatrosses, of Midway Atoll, 62–63,
    64–69, 74, 76
    courting rituals of, 64, 65–68, 69,
        72, 85–86, 89, 175
    juveniles of, 64–68, 72, 85–86, 89,
        175
    nesting season and sites of, 62–63,
        64–65, 68–69, 208–9
    vocalizations of, 65–66, 67, 68, 69,
        85–86
Amazon rain forest, 1–4
    aquatic life in, 13–14
    canopy walkway constructed in,
        2–3, 5, 30, 31
    night sounds and action in, 2–3,
        11, 19–20, 23–25, 26–30, 55
    venomous snakes in, 10–11, 18,
        27

American Sign Language (ASL), 111,
    240, 241
amphibians:
    growth of, 96
    male, and fertilized egg responsi-
        bility, 175
androstenol (Osmone 1), 129
androsterone, 129
anglerfish, 14–15, 116
animal behaviorism, 82, 108, 152, 246
    animal communication and, 54–55
    anthropomorphism rejected in,
        46, 48, 50–53
    changes in direction of, 47–48,
        53–61, 249
    contentious debates in, 34, 38, 41, 44
    cruelty in, 53
    cuing and, 50–53, *51*
    Morgan's canon and, 47–48, 53
    naturalists' approach abandoned
        in, 46–47
    scientists vs. laypersons in, 5, 34
animal cognition, 33
    Darwin on, 43, 250
    Descartes on, 34, 41–43, 44, 45,
        47, 55
    evidence of, 37–39, 40, 41, 50–53,
        54, 55–61, 77, 110–11, 185,
        207, 208, 217, 219, 220, 223,
        228, 232, 234, 238, 244,
        245–47, 249
    see also learning, of animals

animal communication, 3–5, 54
  as behavior manipulation, 9–10
  between species, 5–6, 31, 32,
    35–37, 38–39, 40, 44–45,
    50–53, 82–83, 84–111, 220
  evolution and, 6–7, 8–9, 72–73, 82
  GOP (groans of pain) interpreta-
    tion of, 47–48, 55
  information model of, 5–6, 9, 10,
    112–13
  intentionality in, 38–39, 40
  intention movements and, 25–26
  inventions for studying of, 20, 21,
    146–47, *146*
  natural vs. sexual selection, 8–9
  new science of, 3–4, 19, 33, 55–61
  origins of, 3, 5, 25–26
  senses in, 12–19
  topics of, 32, 185, 249–50
  Wilson's definition of, 6–7
Animal Intelligence (Romanes), 45–46
Animal Mind, The (Morgan), 48–50
Animal Minds (Griffin), 54
animals:
  anthropomorphism and, 34–35,
    36, 45–46, 47, 48, 50–53, 67
  as biological automatons, 34,
    41–43, 44, 45, 47, 55
  language ability of, *see* natural lan-
    guage
  similarities between humans and,
    34, 43, 44–45, 46, 82, 83,
    106–7, 249–50
  tactile stimulation and, 225
  and theory of mind, 41, 44
  tool-making skills of, 37–38
Animal Vocal Communication (Morton
    and Owings), 90
antithesis, in vocal signals and visual
    displays, 98–99
antlers, 167, 168–69
ants, 3–4, 7, 12, 15, 121–23
Ape and the Sushi Master, The (de
    Waal), 71
Apes, Language, and the Human Mind
    (Savage-Rumbaugh, Shanker,
    Taylor), 246
Aristotle, 70

baboons, 168, 186–87, 188
bacteria, 79
  communication between, 5, 6–7,
    112–15, 173
  pathogenic, 6, 113, 114, 116
barks, 87, 89, 90, 91
Bass, Andrew, 138–40
Bassler, Bonnie, 113–14
Bateson, Gregory, 108
bats, 2, 18, 28, 134
  echolocation of, 16, 54, 74–75
  white-lined pheromones of,
    123–24
bees, 8, 121, 122
beetles, Douglas fir, pheromones of,
    124–25
Begemihl, Bruce, 225, 226
Bekoff, Marc, 38, 47–48
Betty (crow), 37–38, 40, 41, 54
Binti-Jua (mountain gorilla), 35–36
biofilms, 113, 114
Biological Exuberance (Begemihl),
    225
bioluminescence, 116–20
birds:
  aggressive territorial displays of,
    74, 89, 91, 92, 97–98, 99,
    150–51, 154, 169, 206–8,
    210–11
  courtship of, 25, 64, 65–68, 74,
    133–34, 135, 145, 149–55, 167,
    168, 169, 170–85
  flight intention signal of, 65, 173
  hatchling begging calls of, 104
  ultraviolet light visible to, 18
  see also songbirds; *specific birds*
birds of paradise, 168, 180–82
birdsong:
  birds dreaming about, 150
  bird vocalizations other than,
    89–90, 91, 92, 155
  learning of, 7, 74, 149–50,
    151–53, 154, 208, 236, 243
  whale song as similar to, 145
Birdwhistell, Ray, 103
blackbirds, 154, 173
black howler monkeys, 193, 194–95,
    196

(black howler monkeys, *cont'd*)
territorial vocalizations of, *56,* 91,
100, 192, 193–96, 206
blister beetles (Spanish fly), 120, 169
blue whales, 16, 145, 148, 159
bonobos, 218, 226, 246–47, *246,* 249
Borgia, Gerald, 185
bowerbirds, 237–38
lekking behavior of, *181,* 182–85
Bradbury, Jack, 123, 129, 132, 134,
151, 155, 168, 179, 180,
199–200, 221
brain, human, cocktail party effect
and, 139
Branham, Marc A., 117–18
Brown, Eleanor, 208
bullfrogs:
female, sounds attractive to,
96–97, 99, 140
vocalizations of, 96–97, 140
buntings, lazuli, 171–73, *171, 172*
butterflies, 121, 167
speckled wood, territorial behavior
of, 196–98, *197*

Cambrian explosion, 115
Canada geese, vocalizations of, 92
canaries, of Midway Atoll, 64
capuchins, brown, 186
cardinal, northern, courtship of, 5
caroling, of Australian magpies, 207
carotenoids, 134, 170–71
Cassano, Ed, 213–14
cats, 15, 89, 123, 225
territorial behavior of, 203, 204
cell phones, as male mate bait, 189
Channel Islands National Marine
Sanctuary, 212–14, 220, 222
chemical signals, 3, 4
as adaptive, 6–7
bacteria and, 5, 6–7, 112–15
territoriality and, 202–4
see also bioluminescence; courting
signals, chemical
Cheney, Dorothy, 55–58
chickadees, black-capped, 150–51
chimpanzees, 11–12, 38, 71, 80,
121–22

aggression and territorial behavior
of, 98, 175, 187, 205–6
courtship and mating of, 170, 175,
187
female, perineal swelling of, 168,
186, 188
gestural communication of,
109–10, 240
greetings of, 224–25
grooming of, 109–10
piloerection in, 98
vocalizations and vocal learning of,
238–40
chinchillas, pheromones of, 123
Chomsky, Noam, 242–43
Chumash culture, 212, 214
Clark, Christopher, 145, 148–49
Clever Hans (horse), 50–53, *51*
coatimundis, 193
cockroaches, Tanzanian, 125
cocktail party effect, 139
cogito, ergo sum, 41
Collins, Sarah, 99–100
colors, 27, 61
courtship and, 167, 169–73, *171,
172,* 180, 203
dominance and, 169–70
communication accommodation,
100–101, 103
conure, orange-fronted, 238
copepods, light attraction of, 28
cosmetics, 189–90
Costa, Daniel P., 17
courting signals, chemical, 112–33
evolution of, 112–16
see also bioluminescence;
pheromones
courting signals, courtship, 107
amplification of, 22
electricity in, 13
general rules in, 136–37
handicap principle in, 134–35,
168–69
honest, 133, 134–35, 168, 180,
186, 188
intention displays in, 173–75
lack of response to, 11, 170

male vs. female investment in, 132–33
nuptial gifts, 120, 174, 177
sound waves and, 226
see also love songs; pheromones; visual courting displays; visual courting displays, human
cowbirds, 7
crickets, amplifiers for, 22
crows, 36, 37–38, 40, 41, 54, 208, 232, 238
cultures:
  animal, 53, 70–72
  and language acquisition, 248

dancing, 166, 173–85
  energetic, 180
  intention movements and, 65, 173–74
  lekking behavior in, 179–85
  male vs. female participation in, 175, 178, 184
  of mammals, 174–75, 167–69, 227
  patterns in, 175–79
Dantzker, Marc, 179
Darwin, Charles, 19, 43–44, 45, 46, 47, 72–73, 77–78, 99, 250
date copying, 71
Davies, N. B., 197–98
Davis, James W., 175–76
Dawkins, Richard, 9–10
Day, Lainy, 183–84
deaf:
  education of, 241–42
  signed language of, 111, 240, 241
deception, 27, 60–61, 133, 169
deoxyribonucleic acid (DNA), 78
Descartes, René, 41–43, 45, 55
Descent of Man, and Selection in Relation to Sex, The (Darwin), 43
de Waal, Frans, 71–72
Dinoponera quadriceps (ant), 122–23
direct benefits model, 168, 169
Discours de la Methode (Descarte), 41
disruptive selection, 171–73, *171*
Dodd, George, 128–29
dogs:
  in Afghanistan, 94–96, 97

aggression and submission displays in, 94–95, 97, 99
communication between humans and, 5–6, 44–45
greeting gestures of, 225
scent marking and surveillance of, 202, 203, 204
vocalizations of, 89, 92–93, 94–95, 203
dolphins, 98, 212–28
  bottle-nosed, 218, 221, 222–24, 225–26
  bubble streams of, 219, 220
  in Chumash legend, 212, 214
  communication organs of, 218
  echolocation used by, 16, 17–18, 215, 217, 218
  human communication with, 220
  learning ability of, 217, 220, 226, 249
  mating of, 226–27
  predictive signaling and, 227–28
  signature whistles of, 5, *213,* 220–24, 230, 240, 243
  sociosexual behavior of, 224–26
  vocal learning and vocalizations of, 5, 16, 21, 213–28, 230, 232, 240, 243
Domb, Leah, 188
domestic selection, 73, 75, 77
dormitory effect, pheromones and, 127
douroucouli (owl monkey), 28–29, 55
Drea, Christine, 48
duetting:
  of birds, 5, 154–55
  of gibbons, 156–57
Dugatkin, Lee, 70–71

ears, and territorial behavior, 204
echolocation:
  of bats, 16, 54, 74–75
  of marine mammals, 16–18, *17,* 215, 217, 218, 233
  in radar and sonar, 16–17
Eckman, Paul, 101
eels, electric fields and, 13–14
eggs, mate selectivity and, 132–33

Eisner, Thomas, 118–20
electrical communication, 13–14
elephants, 204
    contact calls of, 5, 18, 20–21, 157,
        158, 159–60, 161–63, 194
    families of, 157–58, 159, 161
    mating behavior of, 15, 129–30,
        160
    pheromones of, 15, 125–26,
        129–30
    storm and seismic sensing of, 12,
        163–64
emotions:
    expressed similarly in animals and
        humans, 44–45, 46, 82, 83, 99,
        106–7
    stimulated through mother-infant
        communication, 105–7
emotions, animal, 3
    clinical extremism in description
        of, 47–48, 55
    inter-species ability to read, 44–45,
        52–53, 82, 83, 89–90
    theory of mind and, 45, 82
Engstrand, Su, 154
"Ethological Theory on the Origin of
    Vocal Learning, An" (Marker),
    152–53
Everest, Mount, expedition to,
    165–66
evolution, 64
    theory of, 43, 44, 72–76, 82
*Expression of the Emotions in Man and
    Animals, The* (Darwin), 44, 45,
    99
extinction, 76

falcons, dance movements of, 175
Fay, Mike, 163
feathers, courtship and, 134, 135, 167,
    168, 170–73, 178, 180
Fentress, John, 93–94
fer-de-lance snake, 11, 18, 27, 126
Fernald, Ann, 105–7
finback whales:
    courting calls of, 16, 145, 148–49,
        159
    size of, 16, 148–49

finches, 64, 150, 170–71
fire-colored beetle, 169
fireflies, 5, 116–20
fish:
    male, as nest builders, 138–39, 175
    ornamentation of, 138, 167
    schooling, 15
    territorial behavior of, 209–10
    tropical, 171
    see also specific fish
Fisher, Ronald A., 168
Fisherian runaway model, 168
fishing, commercial, 76, 149
flies, balloon, 174
flight intention signal, 65, 173
food calls, 39, 60
Ford, John, 229–31
Fouts, Roger, 110
frogs:
    tree, 2, 3, 22, 29–30
    Tunagra, courting call of, 134
fruit fly, genome of, 79
Full House (Gould), 112

Galápagos Islands, finches of, 64
Gallagher, Timothy, 100, 102–3
geese, 173
genes, genetics, 3, 9, 69–70, 71, 80–82
    conserved, 80–81, 82
    and evolutionary adaptation,
        72–75, 76, 82
    in evolution of human language,
        242–43
    good, mates selected for passing
        along of, 25, 133, 135, 168–69,
        170–73, 180
    and individual variations, 78
    makeup of, 78
    MHC, 130–32
    selfish, 9, 70
Genesis, Book of, 31
genomes, 78–79
    shared by humans and animals,
        79–80, 82
gibbons, duetting of, 156–57
Gilliard, E. Thomas, 182
giraffes, vocalizations of, 18
Givens, David, 107–8, 189–91

Gombe Stream National Preserve
(Tanzania), 98, 206
Goodall, Jane, 98, 109, 206, 216
good genes model, 168–69
Gorak Shep, Nepal, 165–66
Gould, James, 39, 124
Gould, John, *181*
Gouzoules, Harold, 11–12, 58–61,
187
Gouzoules, Sally, 58–61
Grant, Pete and Rosemary, 64
Great Britain, 153–54
cream stolen by birds in, 39–40,
49
Great Chain of Being, 41–43
grebes, great crested, 68, 178
Greene, Erick, 171–73
greeting rumble, of elephants, 161
Gregory, Stanford, Jr., 100–103
Griffin, Donald, 16–17, 54–55, 61
groupers, 209–10
grouses, sage, 166, 179–80, 208
growls, 87, 90, 92–93, 94–95, 97
Guatemala, 192, 193, 201, 206
guinea pigs, pheromones of, 123
guppies, mate choice of, 70–71

hamsters, golden, pheromones of, 126
hand gestures:
in American Sign Language, 111,
240–42
of early humans, 110, 240, 242
of primates, 109–11, 240, 242
handicap principle, ornament evolution and, 135, 168–69
harbor seals, 233, 234–37
Hauser, Mark, 61, 220, 227
hedgehogs, pheromones of, 123
Heinrich, Bernd, 34–35, 36, 232
Helversen, Dagmar and Otto van,
74–75
Hill, David E., 119
Hill, Geoffrey, 170–71
hippopotamuses, vocal signals of, 18
Hoja Bahauddin, Afghanistan, 94
Hölldobler, Bert, 3–4, 7
homeobox genes, 80
honey badger (ratel), honeyguide's

communication with, 38–39,
40, 82
honeyguide (black-throated indicator), 38–39, 40, 82
Hoover (seal), 235–37
hormones, genes for, 80–82
horses, 204–5
courtship of, 174–75
intelligence and perceptiveness of,
50–53, *51*
house sparrows:
aggressive and territorial behavior
of, 92, 97–98, 169, 210–11
vocalizations of, 92, 98
Hrdy, Sarah, 185
Human Genome Project, 79
"Human Maternal Vocalizations to
Infants as Biologically Relevant
Signals (Fernald), 105
humans:
aggression and territoriality of, 98,
201–2, 206
vs. animals after Descartes, 41–43
aural and visual range of, 18
genome of, 79–80, 82
greetings of, 225
mating vs. reproductive cycle of,
188–89
natural economy and, 76, 201–2
piloerection in, 98
hummingbirds, 175, 176
humpback whales:
bubble nets of, 219
songs of, 16, 21, 91, 141–49
hyenas, 48

IMAX, 98
Imitation Factor, The (Dugatkin), 70
immune system, 130
infanticide, 187, 205
infants, human:
and language learning, 7, 8, 57,
153, 236, 244–45, 248
nonverbal communication in,
104–7
infants, nonhuman primate, 58
mother's voice and, 107
insects, 28

(insects, *cont'd*)
"adaptive" communication of, 7
courting rituals and signals of,
116–20, 121–23, 124–25, 174
pheromones of, 121–23, 124–25,
132
Insel, Tom, 81–82
intention movements, 25–26, 173–74
iridescence, in ornamentation, 167
isolation calls, 221

Janik, Vincent, 222–24
Jivaro tribe, 24, 28
Johnson-Sea-Link II, 209–10

Kabul, Afghanistan, 95, 189
Kacelnik, Alex, 37–38
Kanzi (bonobo), 244–45, 246–47, *246,*
249
Katherines, Ginger, 205
Keenan, Julian, 102
killer whales (*Orcinus orca*), 218, 230
bubble nets of, 219
echolocation of, 218, 233
learning ability of, 217, 228,
233–34
vocal learning and vocalizations of,
211, 228–34, 243
kinesics, 108
knife fish, 13–14
Krause, Bernard, 23–24, 74
Krebs, J. R., 9–10
krill, 149

Lamarck, Jean-Baptiste, 77
Lander, Eric, 79–80
language, human, 31–32
artificial, taught to primates, 77,
110–11, 238, 244, 245–47, 249
Biblical story on origins of, 30–31,
248
evolutionary continuity between
human gestures and, 241–42,
243
glottocentric bias in assumptions
about, 240–42
"language gap" theory of, 242–43,
247–48

learning of, 7, 8, 57, 153, 220, 236,
244–45, 248
origins and evolution of, 211, 215,
223, 240, 242–44, 248, 249
see also natural language
Larry King Live, 101, 102
learning, of animals:
by association, 39–40, 49
genes and, 69–70, 71, 152–53
observation and imitation in,
70–72, 152–53, *156,* 184–85,
207–8, 211, 212–48
leks, lekking behavior, 179–85
hot-spot theory of, 179
parties and, 190–91
territorial fights over, 208–9
lemurs, ring-tailed, 204
Lennill, Daniel, 150–51
"let's go" rumble, of elephants, 5,
161–62, 173
Lilly, John C., 215–17, 220
lions, 164, 169–70
lips, 25
as sexual ornament, 189–90
lizards, dance movements of, 175
Lorenz, Conrad, 25, 44
love songs, 138–64
broadcast frequencies of, 140
of songbirds, 25, 74, 134, 135, 145,
149–55, 179–80, 181, 182, 183,
184
Low, Bobbi S., 189–90
lucibufagans, 119
Lupey (wolf), 93
LuxS gene, 113–14
Lyon, Bruce, 171–73

McArthur, 212, 213, 220
macaques, 59–61, 186
Macaulay Library of Natural Sounds
(Cornell University), 19–20
McClintock, Martha, 127, 131
McComb, Karen, 158, 159
McKibben, Jessica, 139–40
Macrides, Foteos, 126
McVay, Scott, 142
Maggot (dog), 95–96

magpies, Australian, 206–8, 238
major histocompatibility complex (MHC), 130–33
Man and Dolphin (Lilly), 215
Marchaterre, Margaret, 138
Margoliash, Daniel, 150
marine mammals, 146
  see also specific animals
Marler, Peter, 59–61, 151–52
martins, purple, 91, 99
Matata (bonobo), 244
Meinwald, Jerrold, 119
Mendel, Gregor, 77
menstrual cycles, queen bee effect and, 127–28
Men's Voices and Women's Choices, 99–100
mice, 126, 130
  genome of, 79–80
midshipman (fish), 138–41
Midway Atoll, 61–69
Milinski, Manfred, 131–32
mimicry:
  of birds, 7, *156*, 207, 232, 237–38
  of chimpanzees, 240
  of marine mammals, 220, 235, 236–37, 240
mind, theory of, 41, 44, 45, 82
Mind and Nature (Bateson), 108
minke whales, vocalizations of, 145
Mir 1, 14
mitochondria, 115, 116
mobbing call, 89
Monitor, 209, 210
monkeys, alarm calls of, 29
  see also specific monkeys and apes
monogamy, 172–73, 190
Moore, Allen, 125
Morgan, C. Lloyd, 47, 48–50, *49*
Morgan's canon, 47–48, 53
Morton, Eugene, 10, 85, 86–91, 103, 154, 199
Moss, Cynthia, 21, 160–62
moths, 28, 75
  pheromones of, 120–21, 132
mountain lions, 35, 36–37, 38
Mucuna, pollinators of, 74–75
musk, 128–29

National Marine Sanctuary Program, 209–10
Native Americans, 166
natural language, 3, 31, 32, 45
  glottocentric bias in assumptions about, 240–41, 242–43
  motivation-structural rules for, 87–88, 90–95, 106, 232
  similarities of human and, 83, 84–111, 176, 214, 223, 243, 248–49
  universal inter-species emotion recognition and, 45, 82, 83
natural selection, 72–78
  adaptation and variation in, 72–75
  for communication and intelligence, 6–7, 8, 10, 31, 54, 82
  misconception about, 75–76
Nature, 131
New England Aquarium (Boston), 220, 236–37
nonverbal communication, human, 101–3, 108, 109
  hand gestures in, 108–9, 110–11, 240–42
  between mothers and infants, 104–7
Norris, Kenneth S., 17, 215, 216, 218
northern right whales, 21, 147

Ober, Carole, 131
Occam's razor, 47
ocean vents, life in, 4, 7, 113
O'Connell-Rodwell, Caitlin, 163–64
Omni, 216–17
On the Occurrence and Significance of Motivation-Structural Rules in Some Bird and Mammal Sounds (Morton), 85, 88
orangutans, 77, 186
Origin of Species, The (Darwin), 43, 72. 76, 77–78
Osten, Herr von, 50–53
Owings, Donald, 10, 90
oxytocin, 81–82
Ozark Mountains, 1–2, 96, 105, 116, 117

Pagel, Mark, 188
paralanguage, 108
parrots, 154–55, *156,* 238, 249
Payne, Katy, 20–21
Payne, Roger, 21, 141–42, 144–45
peacock, 135, 165, 168, 188
Peake, Tom, 150–51
penguins, parent-egg calls of, 68
perfume industry, MHC and, 131–32
perineal swelling, 168, 186, 188
Petén rain forest (Guatemala), 192–96
pet owners, 5, 34
pets, communication with, 84, 104
Pfungst, Oskar, 50–53, 246
pheromones, 19, 120–33
    bacteria and, 113–15
    of insects, 121–23, 124–25, 132
    of mammals, 15, 123, 125–33,
        218
    MHC and, 130–32, 133
    social status and, 125–26, 129
pheromones, human, 123, 125
    male vs. female scents, 128
    and men's musky scents, 128–29
    queen bee effect and, 127–28
    VNO in, 126
piloerection, 26, 98
plastic song, 153
playback studies, 19
polygamy, 190
Poole, Joyce, 21, 160–62
praying mantis, mating of, 175
presidential candidates, U.S., in non-
        verbal communication study,
        102–3
primates, 53
    alarm calls of, 12, 28, 29, 35,
        55–61, *56*
    artificial human language taught
        to, 77, 110–11, 238, 244,
        245–47, 249
    courting displays and mating of,
        168, 175, 185–88, 196
    greeting gestures of, 225–26
    hand gestures of, 240, 242
    infanticide and, 187, 205

territorial behavior of, *56,* 91, 98,
        100, 175, 187, 192, 193–96,
        205–6
    vocal learning of, 238–40
Pygmies, 38–39, 40

queen bee effect, pheromones and,
        127
quorum sensing, 6, 113–14, 173

rain forests, 4, 22–24
    bioacoustic niches in, 23–24,
        73–74
    human damage to, 73–74, 201–2
    sound of, 22, 23, 24
    see also Amazon rain forest; Petén
        rain forest
ravens:
    interspecies communication of,
        35, 36–37, 38, 82–83
    vocalizations and vocal learning of,
        208, 232, 238
Ravens in Winter (Heinrich), 34
receiver (perceiver), 5, 10–12, 112
reporters, 39, 60, 200–201
responses, 5, 6
Reusch, Thorsten, 131
Richards, Whitman A., 175–76
roadrunners, 176–77
rodents, 126, 197
Romanes, George, 45–46, 48, 67
roundworm (*C. elegans*), genome of,
        79
Royal DM70ex organizer alarm,
        210–11
Russell, Michael, 127–28

Sagan, Carl, 112
Santa Cruz Island (Calif.), 212–14
Sapir, Edward, 84
Savage-Rumbaugh, Sue, 244, 246
Schubel, Jerry, 237
sea lions, 232–33
seals, 63, 64, 141, 149
    harbor, 233, 234–37
Seinfeld, Jerry, 10
sender, 5, 10, 112

senses, communication and, 12–19
September 11, 2001, terrorist attack of, 94
sexual selection, 8, 11, 224
  in communication evolution, 8–9, 10, 31
  low-frequency sounds and, 96–97, 99–100
Seyfarth, Robert, 55–58
Shanker, Stuart, 246
Shumaker, Rob, 77
signals, 5, 6, 10, 11–12
  senses and, 12–19
  see also chemical signals; vocalizations, vocal signals
signed language:
  learned by apes, 110–11
  as true language, 111, 240–42
Singer, Alan, 126
skills, inheritance of, 77
Slater, Peter J. B., 224
Smedley, Scott R., 119
Smile of a Dolphin, The (Bekoff), 47–48
Smuts, Barbara, 186
snakes, venomous, 11, 18, 25
sneaker males, 140–41
Snowden, Charles T., 238, 249
Sociobiology (Wilson), 6
sonar, 16–17, 54, 212–13, 215, 217, 218, 233
songbirds, 149–55
  courtship of, 5, 25, 74, 133, 134, 145, 149–55
  dawn chorus of, 23, 149, 154
  declining populations of, 73–74, 153–54
  song-matching contests of, 149–50, 151, 154, 198–99, 208, 210–11
  territorial behavior of, 74, 91, 92, 150–51, 154, 198–99, 206–8, 210–11
  see also birds; birdsong
Songs of the Humpback Whale, 141–42
song sparrows, 151, 198–99
sound fixing and ranging (SOFAR) channel, 16, *17,* 148, 159, 194

sound spectrographs, sound spectrograms, 20–21, 31, *146*
Sound Surveillance System (SOSUS), whale song detected by, 145
Sparky (eel), 13, 14
sparrows, 152–53, 154
  white-crowned, learning ability of, 152–53, 154
  see also house sparrows; song sparrows
sperm, 132
spider, wolf, courtship of, 174
Spong, Paul, 228–29
squid, bobtail, 114
squirrels, 29, 84–86, 89
stickleback fish, MHC and, 131
Stoddard, Philip, 13–14
Stokoe, William C., Jr., 241–42, 243
stripes, 167, 203
Struhsaker, Thomas, 55
Stumpf, Carl, 51–52
submersibles, 14, 209–10
submission, 98–99
subsong, 7, 152–53
survival of the fittest, 78
Swallow, George and Alice, 235–36
sweet talk, 104

Taco (dog), 5–6, 44–45, 203
Taliban, 94, 95, 189
Tarra, Giulio, 241
Taylor, Talbot, 246–48
teeth, baring of, 25
telepathy, with animals, 84–85
termites, pheromones of, 121, 122
territoriality, 192–211
  animal obsession with, 200–201
  female gender and, 205–6
  home owner's advantage in conflicts over, 198, 199–200, 203, 209
testosterone, 129, 170
Theory of the Leisure Class, The (Veblen), 135
Thomas, Roger K., 48
Tikal, 192, 193, 196, 201

Tinbergen, Niko, 25, 44
tit birds, 39–40, 49, 169
Tony (dog), 48–50, *49*
Tower of Babel, 30–31, 248
Train Go Sorry (Cohen), 111
tropicbird, red-tailed, 167, 175, 178
Tyack, Peter, 21, 142, 143, 144–45, 146–47, 218, 230

vasodilators, 15, 120, 129
vasopressin, 81–82
Vehrencamp, Sandra, 123, 129, 132, 134, 151, 180, 199–200
vervet monkeys, 8, 55–58
violence, 25–26
visual courting displays, 165–91
    gender and, 166–68, 185–88, 227
    ornaments, 135, 167–73, 178, 179, 180, 186, 188
    see also dancing
visual courting displays, human, 188–91
    gender and, 188–91
    ornamentation, 189–90
    at parties, 190–91
visual displays, 18
    antithesis in, 99
    see also territoriality
vivisections, 42–43
vocalizations, human, 99–108
    higher frequency in, 88, 103–8
    low frequency in, 88, 99–100, 102–3
vocalizations, vocal signals:
    dissemination and amplification strategies for, 22–24, 159–60
    higher-frequency sounds in, 103–4
    low-frequency, 96–97, 99, 159–60, 194
    motivation-structural rules for, 85, 87–88, 90–95, 99
    parents-offspring, 58, 68, 103–4, 107
    in rain forests, 2–3, 11, 19–20,

22–25, 26–30, 31, *56,* 73–74
    see also love songs; natural language
vocal learning, 211, 212–44
    of birds, 7, *156,* 207, 232, 237–38
    marine mammals and, 211, 212–37, 238, 240
    of primates, 238–40
Voigt, Christian, 123–24
voles, 81–82
vomeronasal organ (VNO), 15, 126–27, 129, 130
Voyages of Doctor Dolittle, 33

Wallace, Alfred Russel, 78
walruses, calls of, 149
warblers, golden, 7
weapons, 167, 168–69, 204–5
Wedekind, Claus, 131–32
whales, *see specific types of whales*
whale songs, 21
    birdsong as similar to, 145
    distances traveled by, 16, *17,* 145, 148, 159–60, 194
    see also humpback whales, songs of
whaling, commercial, 142, 147
whines, 87, 90–91, 93
Whitehead, Hal, 228
Whitten, Patricia, 185
Why Sex Matters (Low), 189
Widder, Edie, 14
widowbird, Jackson's, 177–78
Wilcox, Sherman, 240–41, 243
Wild Boy of Aveyron, 245, 248
Wilson, E. O., 3–4, 6–7, 47
wolves, 93–94, 225
Wrangham, Richard, 239

yaks, 166
Yamazaki, Kunio, 130–31
yellow-naped parrot, 154–55, *156*
Young, Larry, 81–82

Zahavi, Amotz, 135, 168
Zoe (magpie), 208

ABOUT THE AUTHOR

TIM FRIEND is the senior science writer for *USA Today*. During his thirteen-year career at the newspaper, he has covered a broad range of topics including archaeology, anthropology, physics, astronomy, nature, biotechnology, and genetics. He has reported stories from Mount Everest, Antarctica, the Arctic Circle, the Amazon Rain Forest, the Middle East, Central America, and from a one-person submersible on the ocean floor. In addition to his work with *USA Today*, he has written for numerous national magazines, including *National Wildlife* and *Men's Health*. Most recently, he traveled to Afghanistan to report on the war there. He lives in Washington, D.C.